Rethinking the History
of American Education

RETHINKING THE HISTORY OF AMERICAN EDUCATION

Edited by

William J. Reese and John L. Rury

palgrave
macmillan

RETHINKING THE HISTORY OF AMERICAN EDUCATION
Copyright © William J. Reese and John L. Rury, eds., 2008.
All rights reserved. No part of this book may be used or reproduced in
any manner whatsoever without written permission except in the case
of brief quotations embodied in critical articles or reviews.

First published in 2008 by
PALGRAVE MACMILLAN™
175 Fifth Avenue, New York, N.Y. 10010 and
Houndmills, Basingstoke, Hampshire, England RG21 6XS.
Companies and representatives throughout the world.

PALGRAVE MACMILLAN is the global academic imprint of the
Palgrave Macmillan division of St. Martin's Press, LLC and of Palgrave
Macmillan Ltd. Macmillan® is a registered trademark in the United
States, United Kingdom and other countries. Palgrave is a registered
trademark in the European Union and other countries.

ISBN-13: 978-0-230-60009-6
ISBN-10: 0-230-60009-3

Library of Congress Cataloging-in-Publication Data is available from
the Library of Congress.

A catalogue record of the book is available from the British Library.

Jacket photograph courtesy of Herbert M. Kliebard.

Design by Scribe Inc.

First edition: January 2008

10 9 8 7 6 5 4 3 2 1

Printed in the United States of America.

To Carl F. Kaestle, distinguished scholar, mentor, and friend.

CONTENTS

DEDICATION AND ACKNOWLEDGMENTS

This book is dedicated to Carl F. Kaestle, an exceptionally distinguished historian of American education. Conceived in conjunction with his announced plans for retirement, it also marks an apt moment for reflection on the growth and development of his chosen field, which he entered just as it was embracing a decidedly radical and contentious revisionist turn. Carl's early work reflected his concern with such emerging revisionist themes as social and economic inequality, the role of elites in developing school systems, the functions of bureaucracy, and the impact of ideology on education reform. Carl also had strong professional ties to an earlier and somewhat less controversial phase of revisionism in educational history, having studied first with Lawrence Cremin at Teachers College, Columbia University, before moving to Harvard to complete his graduate training with Bernard Bailyn. Unlike his mentors and some of the more radical revisionists, however, he dedicated most of his career to studying schools or closely related subjects such as reading and literacy or educational policy. Carl taught the history of education for over three decades, at several different institutions. His professional life thus spanned a period during which educational history experienced some dramatic changes, as the wheels of historical interpretation turned the field in new directions.

As former students we feel a special debt to our mentor. Along with the other contributors whose work we commissioned, we view this volume as a tribute to Carl's career and scholarly contributions. But this is not a traditional festschrift. It does not focus exclusively or even principally upon Carl Kaestle's work. Rather, it was conceived as a means to think broadly about the state of the field as it emerged over the last few decades. Such a sweeping and challenging task required the cooperation of numerous colleagues and friends, and we were fortunate to enlist a remarkable roster of historians in this enterprise. All of the contributors are friends of Carl; many were his students, others his colleagues. Most have had an affiliation of one sort or another with the University of Wisconsin-Madison, where Carl spent most of his career and trained most of his graduate students. Without question, the range of topics and perspectives found in this volume reflects the field's expansiveness and the diversity of interests that Carl nurtured over the course of his career.

We are grateful to our many colleagues who contributed essays to the book. They wanted to honor Carl, of course, but a broader goal was to provide the next generation of historians of education and other interested readers with a sense of how the field has grown and developed. A number of other people also helped bring this book to publication. Our thanks go to the anonymous reviewers who provided useful suggestions for strengthening the volume. In addition, the staff at Palgrave Macmillan has been a constant source of assistance and encouragement, along with the editors at Scribe Editorial. Most of all, however, we acknowledge especially the critical role of our editor at Palgrave, Amanda Johnson Moon, who has been unflagging in her interest and support for this project from the outset. It is one thing to have an idea, yet another to turn it into a book. Although she left Palgrave just before the book went into production, Amanda's efforts helped in no small measure to make this possible, and for that we are very thankful.

W. J. R. and J. L. R.

CONTRIBUTORS

Jack Dougherty is associate professor of education at Trinity College, Hartford, Connecticut.

Barry M. Franklin is professor of secondary education and adjunct professor of history at Utah State University.

Michael Fultz is professor of educational policy studies at the University of Wisconsin-Madison.

N. Ray Hiner is professor *emeritus* of education and history at the University of Kansas.

Jacqueline Jones is the Truman professor of American civilization at Brandies University.

Gerald F. Moran is professor of history at the University of Michigan, Dearborn.

Margaret A. Nash is associate professor of education at the University of California, Riverside.

Christine A. Ogren is associate professor of education at the University of Iowa.

Michael R. Olneck is professor of educational policy studies and sociology at the University of Wisconsin-Madison.

Adam R. Nelson is associate professor of educational policy studies and history at the University of Wisconsin-Madison.

William J. Reese is the Carl F. Kaestle Professor of educational policy studies and history at the University of Wisconsin-Madison.

John L. Rury is professor of education at the University of Kansas.

Maris A. Vinovskis is the Bentley professor of history and public policy at the University of Michigan, Ann Arbor.

Chapter 1

Introduction: An Evolving and Expanding Field of Study

William J. Reese and John L. Rury

The history of education is an old and venerable field, whose origins as an area of scholarly interest date to at least the early nineteenth century. Like other specialized historical domains, it has experienced interpretive debates and changing schools of thought on a range of issues.[1] The history of American education underwent a major upheaval during the 1960s and 1970s, when a number of scholars challenged long standing views regarding the role of schools in society. While historians had traditionally viewed schools as engines of social and economic development, and as reliable sources of social mobility for every generation of Americans, the so-called revisionist scholars—particularly the "radical" revisionists—argued that the schools reinforced existing patterns of discrimination and inequality. Historically, the schools had delimited and not enhanced opportunity for most children, especially the poor and racial minorities. Not surprisingly, this vividly revisionist interpretation of American educational history proved quite controversial, attracting considerable attention to the field. Not every scholar in the history of education was a revisionist, of course, and every generation, as the saying goes, seems destined to write its own history, which certainly happened in the years that followed the heyday of radical revisionism

in the early 1970s. Since then, educational historians have generated
a new body of scholarship, often asking new questions about the past
in the light of present-day concerns. *Rethinking the History of American Education* offers original essays that highlight post-revisionist
trends, examining research and writing across the large field of historical scholarship on education.

Shaped by the scholarly interests and professional needs of two
vast domains of knowledge—education and history—the history of
education has never been a single entity, its practitioners housed in
schools or departments of education, as well as in history departments. Long a staple of teacher training programs, taught in normal
schools and education departments of universities since the nineteenth century, it became a legitimate subfield of American history
over the course of the twentieth century. This process proceeded
unevenly, in fits and starts. Not until the postwar era, however, did
historians of education become especially concerned about their
standing within the larger historical profession. The year 1960 was a
pivotal time, as Bernard Bailyn published his slender but groundbreaking monograph on colonial education, *Education in the Forming of American Society*. A young Harvard history professor at the
time, Bailyn both offered a stinging critique of the field, and provided a clear demonstration of how to make it more interesting and
relevant to the study of American history.[2]

The publication of Bailyn's book is generally acknowledged as
marking the beginning of "revisionism" in the history of American
education. He called upon the field to view education more expan-
sively, going beyond schooling to study informal processes of acculturation and social development, and to examine such broadly
educative institutions as the family, churches, and the popular press.
The impact was immediate, as other historians soon answered the
call for change. The most influential of these was Lawrence Cremin
at Teachers College, Columbia University, who argued that the history of education should embrace the totality of social and cultural
factors that affect the course of human development.[3] A widely
acclaimed historian himself, Cremin used the Greek term *paedeia* to
denote the vast array of influences that affect human learning in
modern society. His three-volume history of American education,
published over an eighteen-year period, attempted to capture the
richness and variety of educative impulses in the nation's history.[4]
While monumental in scale and generally well received by historians

and educators alike (the second volume won a Pulitzer Prize), Cremin's work also demonstrated the great challenges inherent in Bailyn's charge to conceive of education in such broad terms. If education was thought to include all of the variegated factors that affect human development, should historians abandon their traditional emphasis, the study of schools? Was it possible or desirable to study all the forces that "educate"? What were the boundaries to the field, and how did one distinguish between educative and non-educative influences in society?

Bailyn and Cremin, of course, represented just one strand of revisionism in the history of American education. The other major development that marked a new direction was the appearance of the historians explicitly labeled "revisionist" in the latter 1960s and the 1970s. Those often described as the "radical" revisionists appeared at a time when schools were still seen as an important adjunct of social reform and human progress, as witnessed in the creation of federal programs—such as Elementary and Secondary Education Act (E.S.E.A.) and Head Start—during Lyndon Johnson's "Great Society" initiative, but also as a source of human distress and despair, as documented in best selling tales of the contemporary "urban school crisis." This new expression of historical revisionism in educational studies was initiated in 1968, a volatile moment in modern politics, when Michael B. Katz published *The Irony of Early School Reform*. This wave of revisionism crested five years later in a provocative collection of essays by Clarence Karier, Paul Violas, and Joel Spring—*Roots of Crisis*.[5] The interpretive thread that linked these works was a stark reversal of the dominant themes of progress and enlightened expansion that had characterized the school histories critiqued by Bailyn and Cremin.

Katz and the other revisionist educational historians argued that the public schools were not heralds of freedom and democracy, but had served as instruments of ideological domination and economic exploitation. They emphasized the influence of social and economic elites in the formation of educational policies, underscored the racist and ethnocentric biases of curricula and educational testing, and argued that schools had usually reinforced class inequities and social injustice. Although their conclusions were not universally accepted, and their scholarship became the subject of considerable controversy, the general message was widely influential. Numerous contemporary studies also emphasized themes of social control and inequality, while avoiding some of the revisionists' interpretive flour-ishes.[6] Indeed, the issues raised by radical writers have remained an important element of research and writing in the field.

Obviously, this brand of revisionism and corresponding line of argumentation was quite different from the interpretive tone sug-gested by Bailyn and Cremin in the early 1960s. Rather than expanding the definition of education to include agencies outside of the schools, these revisionists turned attention toward the many unspoken purposes of schools and other institutions, and under-standing how they were affected by larger social and historical forces. In this respect their critique was very much in line with "radical" theoretical perspectives in the sociology of education—even if they were cited inconsistently—and with the "new left" viewpoint then becoming widely influential among many young historians. It reflected a general proclivity to question the motivations of social and political elites and the institutions they controlled, derived in large part from the well-publicized experiences of the recent civil rights movement and student protest era. This historical context accounted for revisionism's origins outside of the field's mainstream traditions, and it explains much of its appeal. It was in step with the times, including the broader interests of activists who wanted to improve schools for disadvantaged children, and historians who sought a more "usable past."[7]

A product of its era, which shaped its rhetoric, passions, and scholarly commitments, the revisionist moment in the historiogra-phy of American education faded into the past by the early 1980s. By 1978, controversy had erupted with the publication of Diane Rav-itch's book, *The Revisionists Revised*, which accused several of the more prominent revisionist authors of shoddy scholarship and of ignoring the many positive contributions of schools in American his-tory.[8] The uproar was, however, relatively brief, as historians also became interested in a new set of issues that energized the field. As early as 1974, David B. Tyack had published *The One Best System*, an influential study of the history of urban schools that seemed to many a more balanced appraisal of the past; it was critical of the failure of schools to live up to their egalitarian promise, but did not condemn schools as harshly as some other historians had, seeing teachers, for example, in a more sympathetic light than many radical revisionists. The publication of David Tyack and Elisabeth Hansot's study of school leaders—*Managers of Virtue* (1982)—also illuminated new trends. Notably, it demonstrated that historians could help explain issues and problems of management and control that had long bedeviled the nation's schools, especially in larger cities. In addition, Carl F. Kaestle's *Pillars of the Republic* (1983) offered a balanced and nuanced account of the common school era, summarizing the best scholarship of previous decades, while drawing upon his own extensive research on the early development of nineteenth-century school systems. In these various works, questions of ideology and social inequality were central concerns. But these scholars did not claim that schools were inherently or fundamentally oppressive institutions, an apparent message in some of the more controversial revisionist accounts.[9]

By the 1980s, a younger generation of scholars, shaped by the revisionists as well as by their critics, labored to make their own contributions to understanding the educational past. A range of studies, drawing upon different approaches to the past, ensured that the field remained eclectic. Social history would lose its appeal before the decade ended, but it proved exhilarating for numerous scholars, wherever they stood on the issues raised by the moderate or radical revisionists. Some scholars embraced the study of the history of childhood and adolescence, and Herbert M. Kliebard wrote a landmark interpretation of the history of competing ideas that shaped the American curriculum.[10] As students of Carl Kaestle, our own work reflected the continuing evolution of the field. Rury and other historians of education studied the interconnections between education, gender, and the labor market.[11] In 1986, Reese examined the effects of voluntary associations and citizen activists in urban school reform in the early twentieth century.[12] Eschewing the wide-ranging topical interests of Bailyn and Cremin, or the ideological stance of the more radical revisionists that followed, many scholars continued to study schools as a way to understand their social functions in the past and influence upon youth.

A remarkable series of award-winning studies appeared in the 1980s and early 1990s, also touching upon some of the themes that revisionists had highlighted, but going well beyond them as well. The first was David Hogan's study of class differences and educational development in Chicago in the early twentieth century—*Class and Reform*—followed by James Anderson's epic account of race and schooling, *The Education of Blacks in the South, 1860–1935*. David Labaree's analysis of school organization and educational credentials in late nineteenth-century Philadelphia—*The Making of an American High School*—appeared in 1988, and Tyack and Hansot's study of gender in education—*Learning Together*—was published in 1990. A few years later, the arrival of Jeffrey Mirel's monumental study of the Detroit Public Schools, *The Rise and Fall of an Urban School System*, marked a decided shift in emphasis within the field. Schools were no longer depicted as sources of exploitation.[13] Like others in the field, Mirel saw these institutions as sites of conflict over a range of issues—matters of interest to a variety of groups in American society. In the language of the day, the schools were part of a "contested terrain."

From the standpoint of scholarship and narration, these works represented a high standard. Each won the American Educational Research Association's biannual book award, and to a great extent dealt with questions that transcended the concerns of the immediate field. In addition to being well written and carefully researched, these books tackled issues big enough to engage a wide range of other scholars concerned with education. They were emblematic of a new turn in the history of American education as a disciplinary subfield. Studies published in the decades following the revisionist era dealt with gender and race, including studies focused both on elementary and secondary schools, and colleges and universities. They also stimulated new lines of research that shed light on hitherto neglected areas of study. In the years following publication of *The Education of Blacks in the South*, for instance, new research focused on African American schooling.[14] At the same time, a wave of studies focused on the education of American Indians also came into view, exemplified by David Adams' award winning book, *Education for Extinction*.[15] Similarly, the publication of *Learning Together* marked the appearance of a range of studies focusing on the history of women's education in the United States, a line of research that has continued to grow up to the present.[16]

The revisionist themes of injustice and exploitation remained important in this broad body of work, but they also were linked to accounts of opportunity and growth. At the same time that schools contributed to inequity, they became sources of hope, enabling those who had suffered oppression to begin realizing their goals and aspirations. This was a new narrative frame of analysis for the history of American education, transcending both the institutionalist-progressive standpoint of the generation that Bailyn had critiqued, and the class dominion and social control perspective of the radical revisionists. It was an interpretive viewpoint that placed processes of change at the center of historical treatments of educational issues. In addition, it spoke to the wide variety of interests that shaped the evolution of schools and allied educational institutions in American history. In short, it represented a new maturity in the field, a willingness to embrace the complexity of education as a social and political process of change, entailing struggle but also growth and the hope of progress. In doing this, it may also have pointed to a new role for the historian in helping to interpret the many contributions of education to the development of American society, as we know it today, as well as a new appreciation for the problems of the present.[17]

SYNOPSIS OF THE BOOK

The essays in this book cover a range of topics in the history of American education, but they only begin to represent the range of questions and scholarly points of view that comprise the field. Like most other domains of historical research and writing, the history of education is continually evolving. Trends in the parent discipline of history inevitably shape the field. Moreover, it is an especially dynamic academic specialty because of its close links to the professional study of education, and such related and controversial questions as educational reform, inequality, and social justice. As suggested earlier, the contributors to this volume have examined a number of traditional subfields within the history of American education. Some of these have been defined by period, such as "the colonial period" or "the age of the common school," while others deal with a particular type of educational institution (for instance, higher education), the education of a particular group (such as women), or a facet of educational practice (as in curriculum history). Most have long been acknowledged as important subfields within the broader community of educational historians, even if some have developed in large part in the past several decades. As such, they have well developed bodies of literature, and the task of summarizing them and identifying critical themes that have emerged since the revisionist era is hardly an easy one. The contribution of the essays collected here, in that case, is one of bringing readers up to date with the literature in each of these sub-specialties, from the end of the revisionist era, roughly the latter 1970s, to the present.

The first of these chapters (Chapter 2) is by Gerald Moran and Maris Vinovskis, and deals with the early development of education in the United States. The authors analyze recent studies of the extent and nature of colonial literacy, as well as the roles of parents and schools in fostering it. They revisit the contentious debates in the late 1960s and 1970s over the spread of antebellum common schools and private elementary schools in light of additional studies in the past twenty-five years. They also assess recent investigations of the rise and development of antebellum public high schools and academies. A wave of new research has carried the field well beyond the controversies of the revisionist period regarding the character of school reform, and has focused attention on such issues as the growth of literacy, African American education, and female schooling. Beyond this, there is the question of education and economic development: just how much did the growth of schooling contribute to the expanding economy during the nineteenth century?

The next chapter (Chapter 3) is a contribution from Jacqueline Jones, who examines the development of black schooling during the Reconstruction era. Other historians of schooling in the Reconstruction South cite the founding of the Savannah Education Association (SEA) as evidence of black people's desire to create and maintain schools within their own communities. Community leaders organized the SEA in January 1865, barely two weeks after General William Tecumseh Sherman made his triumphal entry into the Georgia city. Before too long, Northerners of various stripes—military officials, agents of the Freedmen's Bureau, and missionaries affiliated with the American Missionary Association (AMA)—were expressing dismay over the fact that the black-run SEA seemed determined to hire its own teachers and control its own schools, free of outside influence. In this essay, Jones explores the association's antebellum antecedents, the causes of its downfall within a year and a half, and its lingering legacy, especially among African-American preachers in Savannah. The SEA was a piece of a larger story about Savannah blacks' quest for education and freedom in their schools, churches, social organizations, and family lives, and the roles of white teachers in fostering education. In presenting this intriguing case study, Jones highlights a critical new thread in the history of

African American education: schooling of black children conducted both by white and black teachers.

Michael Fultz expands upon this theme in his examination of historical writing about black education—and particularly black teachers—in the past several decades (Chapter 4). Coinciding with new critical perspectives regarding the reformulation of structural barriers to educational achievement, and social mobility for African Americans in the aftermath of the civil rights movement and its integrationist emphasis, the history of African American teachers has come alive with a range of interpretative analyses. Fultz points to indications of these changing perspectives that can be seen in the early historiography of African American teachers, where the contributions of the black community in providing and supporting its own teachers were recognized from early on. He also suggests that James D. Anderson's path-breaking volume—*The Education of Blacks in the South, 1860–1935*—had a significant impact on subsequent scholarship, along with the work of a number of other important scholars. Another critical interpretive frame is evident in the so-called "good segregated school" literature, providing important correctives to impressions that black schools were uniformly problematic. Additionally, Fultz considers the growth and influence of African American teacher training institutions, demonstrating in an illuminating case study just how new conceptualizations of local activism and conflict can enrich both the institutional and social history of black teachers and their schools.

Moving from the history of African Americans to other ethnic groups in American history, in Chapter 5 Michael Olneck explores the question of immigration and education, a crucial theme in American educational history. This entails a critical assessment of historical scholarship subsequent to the early 1970s, clarifying the role of American public schooling in the lives and prospects of the offspring of European immigrants to the United States in the period between 1890 and 1925. He finds that recent scholarship has produced a complicated picture, in which immigrant families and their offspring negotiated the demands and appropriated the opportunities provided by American schools in ways not necessarily foreseen nor understood at the time. Beyond this, Olneck has capitalized on his familiarity with studies of contemporary immigration to identify continuities and discontinuities in the role of American schools during two very different periods of mass immigration to the United

States. In a pair of "coda" to his original essay, he examines the difficult questions of immigration and the schools today, and considers just what the historical experience of immigration may offer in terms of perspective on these issues.

In the next chapter (Chapter 6), Margaret A. Nash examines the development of women's education in the United States. The scholarship on this question has changed markedly in the last several decades. This was a topic largely ignored by the revisionist historians, and through the 1980s much of the historiography was focused on issues of access to education, highlighting the exclusion of women from institutions of higher learning and secondary education. Nash suggests that there was a tendency to focus on elite schools, but more recently historians have looked to the institutions far more likely to have served women: high schools, academies, seminaries, normal schools, and land grant institutions. Influenced by the fields of gender studies and cultural studies, new work has asked less about access, and more about the meanings attached to education, both by the women who received education and by groups within the broader society. Nash discusses these changes in the historiography of women's education, and situates new research in both the fields of women's history and educational history. In doing this, she places less emphasis on institutions and curricula than on how changes in education impacted the lives of women, reflecting recent interpretive frames in the field.

Next, in Chapter 7, N. Ray Hiner turns to an even broader question in a sweeping introduction to the rapidly growing field of children's history. Recalling his own early impressions as a historian of education, he notes the salience of James Axtell's call for "a waist-high view" of the educational process, one representing the perspective of children. It was an observation imbued with considerable prescience. Since the 1970s, the field of children's history has burgeoned in size and scope, touching upon each of the principal periods in American history and such other major subfields as family history and demographic history, among others. Hiner aptly demonstrates these intersections between various domains of historical scholarship, while describing how the field of children's history has evolved in the past several decades. Noting that the Society for the History of Children and Youth was just established in 2001, and the new *Journal of the History of Children and Youth* is offering its inaugural issue in 2007, he suggests that it is an area of research and writing that has great potential to inform the work of historians of education. His essay highlights the insights to be gained from thinking about the children who lived and learned in the schools and other institutions, adding new layers of complexity to the stories that historians of education have narrated over the years.

Returning to the question of institutions, in Chapter 8, Christine Ogren explores developments within the critical subfield of the history of American higher education. While scholarship in this realm has proliferated in the post-revisionist era, Ogren describes how it has been influenced by the *pre*-revisionist scholars Frederick Rudolph and Laurence Veysey, and by revisionist challenges to their work. She assesses post-revisionist historiography of higher education in four broad categories: sites, students, scholarship, and structures. While most post-revisionist research on the history of higher education has focused on the same sites as earlier research, primarily prestigious colleges and universities, some historians have looked at colleges in less prominent regions and at less traditional types of institutions, such as academies, normal schools, and junior/community colleges. Post-revisionists have also enriched understandings of the history of students' characteristics and aspirations and how they exercised agency to shape their experiences, as well as the history of scholarship, specifically the undergraduate curriculum and faculty research. In addition, historians have examined how internal administrative structures and external legislation, philanthropic foundations, and organizations have shaped higher education, with mixed effects on individual and institutional autonomy. Finally, Ogren briefly suggests that *post*-post revisionist scholarship might focus on the intersections between studies of students, scholarship, and structures, especially across different sites.

In the following chapter (Chapter 9), Barry Franklin looks at the development of his own field of curriculum history in relation to the history of education and to revisionism. Arriving on the scene as a distinct field of study at the beginning of the 1970s, curriculum history developed in tandem with the revisionist movement in American educational history. Franklin suggests that much of the early growth of the field can be seen as a conversation between scholars embracing a revisionist viewpoint and those who challenged the movement's assumptions. Like the broader field of educational history, curriculum history is presently undergoing a process of development that might be called post-revisionism. In particular, he

argues that curriculum historians refined their interpretive orientation in much the same way as influential social historians in the larger
field of educational history. These curriculum historians began to
look more at what was occurring in the schools rather than recommendations concerning what should be taught. Franklin also considers an emerging body of work that employs the perspective of
postmodernism for interpreting curriculum history. He concludes by
exploring what these developments suggest about the state of current curriculum history research and possibilities for the future.

Jack Dougherty next considers the history of education in metropolitan areas during the twentieth century, particularly in recent
decades (Chapter 10). He notes that urban (or metropolitan) history and educational history have been similar as they have developed since the 1970s, but a gap has emerged in recent decades
between two subfields. On one side, metropolitan history has examined how housing and transportation policies, cultural ideals, and
white racial anxieties fueled post-war suburbanization and created
political conditions for urban decline. Yet these works rarely discuss
how schools played a role in this transformation. On the other side,
educational history has traditionally focused on the rise and fall of
urban school districts. But these studies generally do not connect
historical change in urban education with the growth of suburban
schools, and the resulting political and economic shifts in the metropolitan context. Dougherty highlights the need to bridge the historiographic gap between cities, suburbs, and schools, demonstrating
how the two subfields may inform one another. He also suggests
that important insights can be drawn from work in urban sociology
and political science. In the end, Dougherty points out directions for
future research, commenting on several new histories of metropolitan development and education produced by a rising generation of
scholars.

Finally, in Chapter 11, Adam Nelson examines the changing federal role in American education, a topic of seemingly ever-greater
importance at the start of the twenty-first century. He traces the rapidly expanding historiography of the federal role in education since
approximately 1960, touching on topics ranging from aid for the
economically disadvantaged and disabled to racial desegregation and
the rise of standards-based reform and the accountability movement.
Drawing on the work of historians, historical sociologists, and political scientists, Nelson covers the legislative, judicial, and executive

branches, including the rise of the Education Department as a multi-faceted agency, as each has contributed to an evolving federal role in the nation's schools.

In an area of study as wide and deep as the history of education, no single volume can do justice to the many scholars who have contributed to post-revisionist history. Even a cursory examination of the articles and book reviews in the main journals of the field testifies to its expansive nature over the past generation. No survey of any field is ever complete, and an emerging generation of scholars will inevitably ask new questions about the past. As the Dutch historian Pieter Geyl wrote over a half-century ago, history is "an argument without end," and debates about what matters in the present will naturally provoke new interests and understandings of the past.[18] In looking to recent developments in the field, and to the future, it is possible to discuss a number of additional topics and perspectives, the subject of brief commentary in our epilogue. But even this will fail to capture all of the dynamism currently at play in the field. At the same time, *Rethinking the History of American Education* offers readers a convenient window into the last quarter-century of scholarship, presenting essays on major topics from leading scholars. No historian ever gets the last word, and every group of revisionists will be revised.

NOTES

1. For an overview of the early development of the field, see Milton Gaither, *American Educational History Revisited: A Critique of Progress* (New York: Teachers College Press, 2003). For discussion of more recent trends and comparisons to other fields, see Barbara Finkelstein, "Education Historians as Mythmakers," *Review of Research in Education* 18 (1992): 255–97, Ruben Donato and Marvin Lazerson, "New Directions in American Educational History: Problems and Prospects," *Educational Researcher* 29 (Nov 2000): 4–15, and John L. Rury, "The Curious Status of the History of Education: A Parallel Perspective," *History of Education Quarterly* 46 (Winter 2006): 571–98.
2. Bernard Bailyn, *Education in the Forming of American Society: Needs and Opportunities for Study* (Chapel Hill, Published for the Institute of Early American History and Culture at Williamsburg, VA, by the University of North Carolina Press, 1960).
3. For discussion of Cremin's critique of prior studies and his view of the field, see his *The Wonderful World of Ellwood Paterson Cubberley: An Essay on the Historiography of American Education* (New York: Teachers College Press, 1965).
4. Lawrence A. Cremin, *American Education; the Colonial Experience, 1607–1783* (New York: Harper & Row, 1970); Lawrence A. Cremin, *American Education, the National Experience, 1783–1876* (New York: Harper and Row, 1980); Lawrence A.

14   WILLIAM J. REESE AND JOHN L. RURY

Cremin, *American Education, The Metropolitan Experience, 1876–1980* (New York: Harper & Row, 1988).

5. These were among the most widely cited of the "radical revisionist" studies: Michael B. Katz, *The Irony of Early School Reform: Educational Innovation in Mid-Nineteenth Century Massachusetts* (Cambridge, MA: Harvard University Press, 1968); Joel H. Spring, *Education and the Rise of the Corporate State* (Boston: Beacon, 1972); and Clarence J. Karier, Paul C. Violas, and Joel Spring, *Roots of Crisis: American Education in The Twentieth Century* (Chicago: Rand McNally, 1973).

6. For examples of other studies featuring similar themes, see Carl F. Kaestle, *The Evolution of an Urban School System: New York City, 1750–1850* (Cambridge, MA: Harvard University Press, 1973) and Stanley K. Schultz, *The Culture Factory: Boston Public Schools, 1789–1860* (New York: Oxford University Press, 1973).

7. Perhaps the best expression of the "radical" perspective was the controversial book by Samuel Bowles and Herbert Gintis, *Schooling in Capitalist America* (New York: Basic Books, 1976), a largely interpretive and synthetic work that drew heavily upon studies by revisionist historians. For additional contemporary views, see Joseph Featherstone; David Hogan; Mark Stern, "Commentaries," *History of Education Quarterly* 17 (Summer, 1977): 139–158; R. Freeman Butts, "Public Education and Political Community," *History of Education Quarterly* 14 (Summer 1974): 165–183; Jurgen Herbst, "Beyond the Debate over Revisionism: Three Educational Pasts Writ Large," *History of Education Quarterly* 20 (Summer, 1980), 131–45.

8. Diane Ravitch, *The Revisionists Revised: A Critique of the Radical Attack on The Schools* (New York: Basic Books, 1978); also see Jacob B. Michaelsen, "Revision, Bureaucracy, and School Reform: A Critique of Katz," *The School Review* 85 (Feb 1977): 229–46.

9. David B. Tyack, *The One Best System: A History of American Urban Education* (Cambridge, MA: Harvard University Press, 1974); David Tyack and Elisabeth Hansot, *Managers of Virtue: Public School Leadership in America, 1820–1980* (New York: Basic Books, 1982); Carl F. Kaestle, *Pillars of the Republic: Common Schools and American Society, 1780–1860* (New York: Hill and Wang, 1983).

10. See the essays in N. Ray Hiner and Joseph M. Hawes, eds., *Growing Up in America: Children in Historical Perspective* (Urbana: University of Illinois Press, 1985); also see Joseph M. Hawes and N. Ray Hiner, eds., *American Childhood: A Research Guide and Historical Handbook* (Westport, CT: Greenwood Press, 1985); and Herbert M. Kliebard, *The Struggle for the American Curriculum, 1893–1958* (Boston: Routledge & Kegan Paul, 1986).

11. For this line of inquiry, see John L. Rury, *Education and Women's Work: Female Schooling and the Division of Labor in Urban America, 1870–1930* (Albany: State University of New York Press, 1991), Jane Bernard Powers, *The "Girl Question" in Education: Vocational Education for Young Women in the Progressive Era* (Washington, DC: Falmer, 1992), and Karen Graves, *Girls' Schooling during the Progressive Era: From Female Scholar to Domesticated Citizen* (New York: Garland, 1998). A related line of research and writing focused on the relationship between education and work; see Harvey Kantor, *Learning to Earn: School, Work and Vocational Reform in California, 1880–1930* (Madison: University of Wisconsin Press, 1988) and Herbert M. Kliebard, *Schooled to Work: Vocationalism and the American Curriculum, 1876–1946* (New York: Teachers College Press, 1999).

12. This reference is to William J. Reese, *Power and the Promise of School Reform: Grassroots Movements during the Progressive Era* (Boston: Routledge & Kegan Paul, 1986), one of the relatively few post-revisionist studies to focus on progressive school reform. For research in a somewhat similar vein, see Ronald D. Cohen, *Children of the Mill:*

Schooling and Society in Gary, Indiana, 1906–1960 (Bloomington: Indiana University Press, 1990). Also see the essays in Susan F. Semel and Alan R. Sadovnik, eds. *Schools of Tomorrow, Schools of Today: What Happened to Progressive Education* (New York: Peter Lang Publishing, 1998), and William J. Reese, "The Origins of Progressive Education," *History of Education Quarterly* 41 (Spring 2001): 1–24.

13. David John Hogan, *Class and Reform: School and Society in Chicago, 1880–1930* (Philadelphia: University of Pennsylvania Press, 1985), yet another re-examination of progressive reform; James D. Anderson, *The Education of Blacks in the South, 1860–1935* (Chapel Hill: University of North Carolina Press, 1988); David F. Labaree, *The Making of an American High School: The Credentials Market and the Central High School of Philadelphia, 1838–1939* (New Haven: Yale University Press, 1988); David Tyack and Elisabeth Hansot, *Learning Together: A History of Coeducation in American Schools* (New Haven: Yale University Press, 1990); and Jeffrey Mirel, *The Rise and Fall of an Urban School System: Detroit, 1907–81* (Ann Arbor: University of Michigan Press, 1993).

14. See, for instance, Vanessa Siddle Walker's prize-winning study, *Their Highest Potential: An African American School Community in the Segregated South* (Chapel Hill: University of North Carolina Press, 1996), Eric Anderson and Alfred A. Moss, Jr, *Dangerous Donations: Northern Philanthropy and Southern Black Education, 1902–1930* (Columbia, MO: University of Missouri Press, 1999), Jack Dougherty's award-winning book, *More Than One Struggle: The Evolution of Black School Reform in Milwaukee* (Chapel Hill: University of North Carolina Press, 2004), and Heather Andrea Williams, *Self-Taught: African American Education in Slavery and Freedom* (Chapel Hill: University of North Carolina Press, 2005), also an award winner. Yet other studies are cited in the essays by Jacqueline Jones and Michael Fultz included in this volume.

15. David Wallace Adams, *Education for Extinction: American Indians and the Boarding-School Experience, 1875–1928* (Lawrence: University Press of Kansas, 1995). Adam's study demonstrated the brutality of the boarding school regime, but also the complexity of the response by students, parents and the larger American Indian community. Other studies suggested that assimilation sometimes was a goal of tribal leaders, even in institutions under native people's control. Themes of discrimination, exclusion and manipulation were hardly neglected, but a new emphasis was evident in focusing the beliefs and goals of the schools' clients. In this respect, the emerging scholarship on American Indian schooling has come to exhibit a range of perspectives quite parallel to the larger field of American educational history. Indeed, it is possible to say that it has been among the most vital and visible manifestations of a growing sophistication in the interpretive frames employed by historians of education in the United States. See, for instance, Devon A. Mihesuah, *Cultivating the Rosebuds: The Education of Women at the Cherokee Female Seminary, 1851–1909* (Urbana: University of Illinois Press, 1993), Clyde Ellis, *To Change Them Forever: Indian Education at the Rainy Mountain Boarding School, 1893–1920.* (Norman: University of Oklahoma Press, 1996), Michael C. Coleman, *American Indian Children at School, 1850–1930* (Jackson: University Press of Mississippi, 1993), Amanda J. Cobb, *Listening to Our Grandmothers' Stories: The Bloomfield Academy for Chickasaw Females, 1852–1949* (Lincoln: University of Nebraska Press, 2000), Margaret Connell Szasz, *Indian Education in the American Colonies, 1607–1783* (Albuquerque: University of New Mexico Press, 1988), and Devon A. Mihesuah, *Natives and Academics: Researching and Writing about American Indians* (Lincoln: University of Nebraska Press, 1998). For a recent treatment of American Indian life, focusing on questions of socialization and assimilation in the colonial period, see Amy C. Schutt, *Peoples of the River Valleys: The*

Odyssey of the Delaware Indians (Philadelphia: University of Pennsylvania Press, 2007).

16. See the studies listed in note 11 above, and the award-winning book by Jane H. Hunter, *How Young Ladies Became Girls: The Victorian Origins of American Girlhood* (New Haven: Yale University Press, 2002). Additional studies are discussed in the essay by Margaret Nash in this volume.

17. For two exceptional models of such "post-revisionist" scholarship, undertaken by teams led by prominent historians of education, and tackling issues of considerable contemporary significance, see David Tyack, Robert Lowe, and Elisabeth Hansot, *Public Schools in Hard Times: The Great Depression and Recent Years* (Cambridge, MA: Harvard University Press, 1984) and Carl F. Kaestle, Helen Damon-Moore, Lawrence C. Stedman, and Katherine Tinsley, *Literacy in the United States: Readers and Reading since 1880* (New Haven: Yale University Press, 1991).

18. Pieter Geyl, *Napoleon: For and Against*, translated from the Dutch by Olive Renier (New Haven and London: Yale University Press, 1949), 16.

CHAPTER 2

LITERACY, COMMON SCHOOLS, AND HIGH SCHOOLS IN COLONIAL AND ANTEBELLUM AMERICA

Gerald F. Moran and Maris A. Vinovskis

The study of colonial and antebellum American education received a major impetus almost five decades ago when Bernard Bailyn and Lawrence Cremin challenged scholars to critically re-examine education and schooling in the past (especially employing a broader definition of education than had been used by most previous authors).[1] Since the mid-1960s scholars have made major contributions to our understanding of the role of parents, churches, and schools in educating early Americans. Yet there has been relatively little overlap between the historians who investigate education before and after 1800. In addition, the education issues addressed by colonial historians differ from those pursued by antebellum analysts—partly reflecting societal variations in the past as well as the particular concerns of the current scholars studying those two timeperiods.

In this chapter, we focus on a few topics of special interest to scholars studying colonial and antebellum education and schooling. First, we analyze recent studies of the extent and nature of colonial literacy, as well as the roles of parents and schools in fostering it. Then we revisit the often-contentious debates of the late 1960s and

1970s over the spread of antebellum common and private schools, drawing upon subsequent studies produced during the past thirty years. Finally, we discuss the origins and development of antebellum public schools and private academies.

COLONIAL LITERACY, EDUCATION, AND SCHOOLING

The current state of early American educational history was influenced to a large degree by historians' enthusiastic response to themes contained in Bernard Bailyn's classic work, *Education in the Forming of American Society.*[2] Taking a cue from Bailyn's emphasis on education as a broad social-cultural process involving families, churches, and communities—as well as schools—historians have produced considerable work on the cultural roles of hitherto neglected historical agents, such as children, women, parents, and church laity in the formative period of American history.[3]

At the same time, however, some recent social-cultural histories continue to minimize or ignore the contribution of education to colonial American social and economic development. Even monographs on colonial education sometimes have not adequately explored the possible impact of schooling on the development of early American literacy.[4] As a result, historians trying to link the new social-cultural history to education find themselves in the unfortunate position of lacking adequate studies of formal schooling to do so, as noted by Joel Perlmann, Silvana Siddali, and Keith Whitescarver:

> Before the 1960s, American educational historians routinely produced histories of colonial schools—although the gender-related aspects of the topic were never the central focus for these earlier historians. With Bernard Bailyn's celebrated critique of this earlier research, and with the changing priorities established by the revisionists and others in the 1960s and 1970s, it seemed that only an intellectual slouch would care about the institutional history of colonial schools. Now we find ourselves in the curious situation of needing a better understanding of colonial institutional realities in order to answer the questions of our own time.[5]

In developing a deeper understanding of the institutional context of colonial education, we might look at the work of early modern

European educational historians. These scholars never stopped studying schools; instead, they brought the history of these institutions into line with other educational phenomena, especially the development of popular literacy. Since the 1950s, such historians— using both "indirect" indices of literacy (including the number of schools, the production and sale of books, and inventories of probated estates), and more "direct" measures of literacy (namely the ability to sign a name on a document)—have amassed considerable data on rates of popular literacy.[6]

Accessible through recent syntheses, especially R. A. Houston's *Literacy in Early Modern Europe: Culture and Education 1500–1800,* this work reaches several key conclusions regarding literacy and its social-institutional milieu. One has to do with the trend toward "mass literacy" during the eighteenth century in Europe: "By 1800 over most of north-western Europe, more than half of adult males could sign their names and still more could read a simple text. Whole social groups had command of high levels of functional skills— reading, writing, counting and languages—and there was no class wholly devoid of literate elements. Female literacy advanced less rapidly and did not reach the level of males as a group. Nevertheless, countries like England, lowland Scotland, the Scandinavian cluster, much of Germany and north-eastern France could, in their own distinctive ways, boast mass literacy which set them apart from much of the Mediterranean world and vast areas of central and eastern Europe."[7]

A second significant finding is that the primary source of mass literacy was schooling. To be sure, literacy was "acquired by following a variety of paths," informal as well as formal.[8] Wherever the supply of elementary schools matched the popular demand for them, movement toward mass literacy took place. As Houston shows, this was especially the case in most of early modern northwestern Europe (the home, significantly enough, of most emigrants to British North America). A rising desire for learning, fueled by broad popular movements like the Reformation, Counter Reformation, and Enlightenment, combined with "a remarkable rise in documented schooling" to produce popular literacy.[9] Marked surges in available schooling were often sparked by national literacy campaigns, in which church and state worked in tandem, both to push schooling and to raise popular interest in reading and writing, especially for religious purposes. As Houston says, "co-operation between ecclesiastical and secular authorities in the creation of national educational systems and the fostering of literacy was central to the success of initiatives before the eighteenth century."[10]

Thus, in lowland Scotland, the Calvinist Reformation produced increasing popular interest in reading; but it was not until the Scottish parliament started legislating local schooling into law during the seventeenth century that widespread literacy surged. When supplemented by home schooling of biblical and catechetical readings, the campaign produced near total literacy by the early eighteenth century.[11]

Yet, literacy campaigns were also waged in the eighteenth century, albeit often in more secular contexts, where changes in material conditions such as rising population, urbanization, and expanding commerce could spur popular demand for literacy. But as Houston argues, "the spread of secularism" should not "be exaggerated," and even in the eighteenth century the church provided many of the teachers and much of the drive behind day-to-day schooling."[12] An additional push and pull toward literacy were provided by the revival of religion and the rise of pietism at the end of the seventeenth century. Even as pietistic preachers encouraged their parishioners to acquire basic literacy skills, pietistic lay people "used reading and writing in their search for a satisfying relationship with God."[13]

The earlier work of European literacy historians was well known to early American advocates of the "new" social history, and was incorporated into studies of colonial education as early as 1974, when Kenneth Lockridge published his pioneering work, *Literacy in Colonial New England*.[14] Influenced especially by the research of Swedish literacy historians, Lockridge used European techniques of mark-signature analysis to trace patterns of literacy in colonial New England. He discovered that, in certain respects, New England's educational history echoed developments in contemporary northwestern Europe. New England males achieved near universal literacy toward the end of the colonial period. Females also experienced gains in literacy, but nevertheless lagged well behind the men by the end of the eighteenth century.

As in Europe, the key source in the rise in male literacy was schooling. But Lockridge found no national campaign waged by church and state behind the surge in literacy, although "intense Protestantism" combined with a "concern for education" expressed in "the famous school laws" explained the provision of "systematic schooling" by the Puritans of New England.[15] Rather, literacy did not take off until certain socio-demographic forces had kicked in. Only when rapid population growth had produced "increasing social concentration did these schools become more readily available, hence more effective in raising male literacy."[16] Given the importance of public schools as a source of literacy, it appeared likely to Lockridge that any failure of women to achieve educational parity with men was a product of public school discrimination against female education.

Despite the importance of its findings, *Literacy in Colonial New England* stimulated surprisingly few successors. Instead, some historians immediately challenged Lockridge's use of wills and mark-signatures on wills to measure literacy. Their main objections were that wills, because of the physical infirmities of testators, under-recorded literacy, and mark-signatures, because of their impreciseness, shed little light on the actual educational attainments of signers versus markers. At the same time, with the "turn" toward cultural and linguistic history, early American educational historians shifted interest from quantitative methodologies to anthropological and linguistic models of human experience, including education.

Historians are nevertheless returning slowly to the issues raised in the Lockridge volume and to the work of European literacy scholars. Perhaps the most important line of inquiry has to do with the issue of when, where, and why New England women acquired mass literacy. In an important study of colonial literacy, Linda Auwers tested Lockridge's findings regarding the persistence of female illiteracy by tracking female mark-signatures in the deeds of colonial Windsor, Connecticut. Expecting to find stagnating rates of female literacy for the eighteenth century, Auwers was surprised to discover that Windsor women in fact achieved near universal literacy by the end of the colonial period.[17] Auwers also found that this "dramatic transition in female education," this "revolution in which—virtually all women acquired elementary skills in reading and writing" was produced by schools that were far more inclusive than Lockridge realized.[18] As Auwers concludes: "The interaction between Edwards and his female parishioners in producing higher literacy rates for girls remains speculative, but improved educational opportunity was a potential social effect of female church membership."[19]

While the Auwers article sparked little initial response, new work on the issue of New England women and mass literacy eventually appeared. In contrast to the work of E. Jennifer Monaghan and David Hall, which argues that women as well as men achieved near universal reading literacy as early as the end of the seventeenth century, several recent Massachusetts mark-signature studies echo Auwers in dating the rise of rural female literacy in the period after 1750 (but falling short of the high level reached by Windsor women born between 1730 and 1749). Gloria Main's studies of female signing rates on deeds in rural and urban Massachusetts, for example, found substantial increases by the mid-eighteenth century; and Boston women continued to have higher rates of mark-signature literacy than their rural counterparts prior to the American Revolution. William Gilmore's analysis of deeds from a rural Vermont community points to the 1790s as a crucial decade.[20]

In the most inclusive study of female literacy in New England to date, Joel Perlmann and Dennis Shirley used retrospective evidence on New England women above age seventy listed in the 1850 U.S. census (which was the first census to record individual respondent's literacy). They found that nearly all New England women born after 1765 were literate, a finding in line with Auwers' data.[21] In a subsequent study, Perlmann joined with Silvana Siddali and Keith Whitescarver to explore the question of why women achieved literacy when they did. Drawing on the work of Walter Herbert Small—*Early New England Schools*, published nearly a century ago—and on the records of fourteen Massachusetts towns, they found evidence compatible with the "social concentration" explanation for educational development: changes in schooling linked to the "process of population dispersion and accompanying demands for services in the outlying areas" paved the way for rising female literacy. But Perlmann and his colleagues went beyond Lockridge and Auwers to identify another stage in the process of educational change: the hiring by towns of women to teach reading, and then, gradually, to teach writing. As the authors conclude, "considerable change was, in fact, occurring earlier than 1760—longer school attendance for girls in lower schools, evolving teaching roles for women, and increasing literacy as measured by signing ability."[22]

While we now have some useful ideas about the possible general institutional trends behind increases in women's literacy (but ones that remain to be tested), we still lack an adequate understanding of how girls' schooling and literacy varied across time and space. Kathryn Kish Sklar, in her important article on Massachusetts schools and female students in the period 1750–1820, found that communities differed markedly in their attitudes toward, and approach to, the schooling of girls. Sklar suggested that community gender education inequities "diminished more quickly and easily when male elites were relatively weak, when most of the town's wealth resided in the middle levels of wealth distribution, and when religious authority was decentralized."[23]

Scholars working on the colonial gender differences in literacy and schooling are quick to point out the need for additional research. Sklar, for example, acknowledged that her sample of towns may not be representative; therefore, similar investigations should be undertaken at other sites to determine whether the relationships she found between community characteristics and gender education inequality occurred elsewhere.[24] Main called for future studies of how women acquired literacy—as children in schools or as adults through self-help activities.[25] Perlmann and his coauthors suggested the need for additional empirical studies of the transitional period 1770–1830, more information on the methods of education used to produce literacy gains, further linkage of literacy data to specific birth cohorts, and more mark-signature data for which the age at signing was available.[26]

There is also a troubling tendency among early educational historians to view New England as Massachusetts writ large. Though published nearly thirty years ago, the Auwers study has yet to be followed up with additional data on other Connecticut communities. Was Windsor normative? If so, was Connecticut decidedly different from Massachusetts in the speed with which it advanced toward mass literacy, and if so, why?

A simple political difference between the two colonies raises the possibility of their moving in radically different educational directions over the course of the eighteenth century. In 1692, Massachusetts—in the aftermath of the Glorious Revolution—became a royal colony. Connecticut, however, managed to retain its charter, and thus hang on to its relative autonomy from England. As soon as Massachusetts became a royal colony, it sought to restore all former compulsory education acts, but was prevented from doing so by the disallowance of the statute when it came under review in England. For the rest of the colonial period, the Massachusetts legislature refrained from enforcing mandatory legislation for all children.[27] On the other hand, the Connecticut assembly, free of English legal restraints, enacted more educational legislation than any of the other thirteen colonies.[28] How did these developments affect patterns of education and literacy in the two colonies? Ignorance of the colonial Connecticut education situation prevents us from answering this and other important comparative questions regarding variations in educational experiences throughout New England.

Some research on the history of schooling in colonial Connecticut in relationship to educational policy and mass literacy is presently underway. Examining the extant parish, town, church, and legislative records for the colony of Connecticut, Moran and Vinovskis have found that throughout the eighteenth century, the Connecticut legislature consistently pushed education, in part by adopting—as early as 1700—a school fund to help subsidize schooling at the parish level. At the same time, the Connecticut legislature, intent on maintaining church support for education, shifted responsibility for schooling from the town to the ecclesiastical parish. Meanwhile, the feminization of church membership improved attitudes toward female education, while the Great Awakening, by injecting European-style pietism into the society, reinvigorated interest in teaching and reading the Bible. The result was that levels of male and female literacy in communities such as Windsor and Middletown, Connecticut exceeded those in contemporary Massachusetts before the American Revolution.[29]

Just as we need more studies of areas of New England other than Massachusetts, we also need comparative analyses of New England and European educational and literacy patterns. Although Lockridge pioneered this effort over thirty years ago, early American historians seem reluctant to undertake such studies, although this would certainly stimulate new questions and additional work in the field. Placing present understanding of New England educational history within the context of information found in the Houston volume suggests several possible lines of future inquiry. These include comparative analyses of the pathways to mass literacy in the various areas of New England and Europe. To what extent was cooperation between state and church behind surges toward universal literacy? What was the relative impact of push versus pull factors in producing elevated literacy levels? What was the role of schooling versus informal institutions in popular literacy movements? To what extent were national, interregional, or regional literacy campaigns behind the surges in education? These and other important questions await further analyses.

Finally, there is a need to study education and literacy in regions of British North America outside of New England. Surprisingly little is presently known about literacy and education in the middle Atlantic and southern regions of the British colonies. An important exception is Farley Grubb's analysis of education clauses in the contracts of German immigrant children in Pennsylvania, which found a marked increase in employer-provided education in formal schools between the 1770s and the 1790s, with few gender differences in education.[30] In addition, Robert Gallman's study of literacy in colonial Perquimans County, North Carolina, found a sharp rise in mark-signature rates among free males in areas undergoing rapid social concentration, to use a phrase coined by Lockridge.[31] As both of these studies indicate, the level of education among immigrant populations may have played a far greater role in the educational development of the eighteenth-century mid-Atlantic and South, where immigration rates were higher than in New England, but such determinations must await further research.[32]

There was one sizeable immigrant group that, when it came to literacy, started from scratch upon arrival to America. As recent analyses of literacy and slavery have demonstrated, the thousands of Africans entering colonial America, primarily the South, had no system of writing and had never encountered literacy before entering the colonies.[33] These populations were the subject, however, of literacy campaigns that, to the degree they stressed formal instruction for church-related purposes, appear comparable to the literacy campaigns of contemporary Europe. Indeed, the reliance on schooling by the English Society for the Propagation of the Gospel (S.P.G.) in their campaigns to educate Native as well as African Americans reveals the high value placed on that institution by English colonizers. So, too, does the dependency on schooling among Southern revivalists, who waged war against the illiteracy of poor whites and slaves during the course of the Great Awakening.[34]

It would be important to determine, if only for comparative purposes, the effectiveness of such efforts. While mark-signature data are not available for most African immigrants and their descendants, indirect measures of literacy contained, for example, in the records of the S.P.G. or of the revivalists reveal the speed with which educational values spread through African American communities.[35] If, as

Jill Lepore has recently argued, Native Americans suffered for their literacy, the same cannot be said for African Americans, who saw in the acquisition of reading and writing skills a route to freedom.[36] Of the many issues awaiting future research, comparative analysis of the importance of schooling and literacy to various ethnic, religious, and regional groups in British North American is surely among the most enticing.

ANTEBELLUM COMMON SCHOOLS

A number of studies of antebellum elementary or common schools had appeared before the 1960s.[37] Some were broad overviews in textbooks about public education by authors such as Ellwood P. Cubberley.[38] Others were monographs on states such as Massachusetts and New York.[39] Articles and books also had been published on specific aspects of the common schools, such as rural education and elementary teachers.[40] Particular attention was focused on the reform activities of Horace Mann, the first Secretary of the Massachusetts Board of Education.[41]

Many of the studies of antebellum common schools were written by professional educators who praised the progress made by public schools during these years. In the 1960s, Bernard Bailyn and Lawrence Cremin criticized these earlier works as lacking an appropriate historical context, downplaying public school shortcomings, or narrowly conceptualizing education by ignoring the impact of other factors, such as parents, schools, and the popular media. Despite their misgivings about the scope and interpretations of these earlier studies, Bailyn and Cremin praised school reformers such as Mann, and acknowledged the valuable, though limited, contributions of schools.[42]

Reacting to the 1960s urban school crisis, a new generation of scholars rejected traditional explanations of the origins of antebellum common schools, and questioned the education provided by these institutions. The so-called "revisionist" historians were a small, diverse, and influential group of critics. In the late 1960s and early 1970s, these analysts stimulated intense, often contentious, debates about common schools. Other useful investigations of antebellum education that did not focus directly on the issues raised by the revisionists were produced; yet these studies often received less scholarly and public notice at that time.[43]

Michael Katz, one of the earliest and most active revisionist historians, in his 1968 groundbreaking monograph, *Irony of Early School Reform*, rejected earlier interpretations of antebellum Massachusetts education: "Americans share a warm and comforting myth about the origins of popular education. For the most part historians have helped to perpetuate this essentially noble story, which portrays a national, enlightened working class, led by idealistic and humanitarian intellectuals, triumphantly wresting free public education from a selfish, wealthy elite and from bigoted proponents of orthodox religion."[44]

Katz compared the mid-nineteenth-century Massachusetts reforms with those advocated by James Bryant Conant and others in the 1960s. He saw both reform efforts "spearheaded by the socially and intellectually prominent concerned for the preservation of domestic tranquillity and an ordered, cohesive society. In both cases this group has been joined and supported by middle-class parents anxious about the status of their children and, somewhat tardily, for the most part, by the organized schoolmen, who understandably enough have usually evaluated reform theory in terms of its impact upon their own precarious status." Katz added that "very largely both movements of urban reform have been impositions; communal leaders have mounted an ideological and noisy campaign to sell education to an often skeptical, sometimes hostile, and usually uncomprehending working class. The reformers have been determined to foist their innovations upon the community, if necessary by force."[45]

While readers sometimes interpret the *Irony of Early School Reform* as including an analysis of Mann's common school initiatives, the three case studies Katz investigated focused more narrowly on the abolition of the Beverly High School in 1860, the attack by the American Institute of Instruction on Cyrus Pierce, and the debates over juvenile delinquency and the Massachusetts state reform school.[46]

Other revisionists investigated the antebellum Massachusetts common schools more directly. In the mid-1970s, Samuel Bowles and Herbert Gintis, for example, drew upon studies by Katz and other scholars to argue that manufacturers and merchants, allied with professionals, spearheaded the education reform movements in order to develop a more disciplined labor force:

> There can be little doubt that educational reform and expansion in the nineteenth century was associated with the growing ascendancy of the capitalist mode of production. Particularly striking is the recurring pattern of capital accumulation in the dynamic advanced sectors of the economy, the resulting integration of new workers into the wage-labor system, the expansion of the proletariat and the reserve army, social unrest and the emergence of political protest movements, and the development of movements for educational expansion and reform. We also find a recurring pattern of political and financial support for educational change: While the impetus for educational reform sometimes came from disgruntled farmers or workers, the leadership of the movements—which succeeded in stamping its unmistakable imprint on the form and direction of educational inno- vation—was without exception in the hands of a coalition of profes- sionals and capitalists from the leading sectors of the economy.[47]

As revisionists put forth their re-interpretations of antebellum education reforms, historians such as Diane Ravitch challenged their work. Calling for a more balanced portrayal of American education, Ravitch questioned the revisionists' interpretations of the motiva- tions of reformers such as Mann and rejected the dismissal of any relationship between education and social mobility in the past.[48] The revisionists staunchly defended their analyses and questioned why the National Academy of Education had commissioned Ravitch to critique their work.[49] This exchange in the late 1970s reflected not only scholarly differences, but also revealed the strong, often acri- monious debates about education in the past and present.[50]

Several important studies of antebellum urban schooling were completed in the early 1970s. Carl Kaestle, for example, analyzed New York City schools from 1750 to 1850. Most of these studies did not provide an in-depth analysis of U.S. school attendance.[51] In the mid- 1960s, however, Albert Fishlow published a challenging, though often overlooked essay on antebellum schooling. Fishlow argued that Mann's reforms had not significantly increased Massachusetts school attendance. Instead, the major increases in antebellum schooling between 1840 and 1860 occurred in the north-central region rather than in New England or the Middle Atlantic states.[52]

Comparable findings were reported by Kaestle and Vinovskis in their statistical analysis of Massachusetts education, as well as by Lee Soltow and Edward Stevens in their study of national literacy and schooling.[53] Thus, contrary to the arguments of Bowles and Gintis,

Massachusetts school attendance was already high in 1800, well before the arrival of Mann or the industrialization of the Commonwealth.[54] Moreover, the areas which experienced the largest gains in school attendance in the two decades prior to the Civil War remained largely rural and agricultural, rather than regions experiencing the most industrial development.[55]

Despite impressive gains in school attendance and literacy outside of New England and the Middle Atlantic states, large regional differences remained on the eve of the Civil War. While opportunities to attend schools in the South increased, the region trailed the rest of the nation. Additionally, large numbers of Southern white women continued to be illiterate even as most Northern white women attended schools and learned to read and write.[56]

Many industrialists and merchants supported common school reforms, but they were not necessarily the key proponents, especially in more rural and agricultural states such as Michigan and Ohio. Others, such as clergymen and Whig or Republican political leaders, also played important roles in advocating antebellum school reforms, and Northern workers often supported common schools and sent their children to those institutions.[57]

Recent interpretations of the rise of antebellum common schooling often point to the importance of the general Protestant emphasis on everyone being able to read the Bible, as well as the need for a better-educated white male electorate in the new republic. At the same time, growing commercialization and expansion of the market economy reinforced the value of an educated labor force. The growing availability of newspapers and novels, as well as the expanded circulation of personal letters, also reflected and benefited the increasingly literate population.[58]

The attention of the revisionists and their critics was mainly on the antebellum public schools. Today, however, there is growing interest in the role of private academies at both the elementary and secondary levels.[59] There is also more attention given to the education of girls in the public common schools, as well as their subsequent job opportunities as school teachers. Indeed, in a few areas such as pre-Civil War Massachusetts, an estimated one out of five white women briefly taught school at some point in their lives.[60]

Although most of the focus of education historians has been on antebellum whites, there has been additional work on the education of free African Americans and slaves. Contrary to the assertions of sociologists such as John Ogbu, free and slave African Americans were very interested in education before the Civil War. Despite the outright hostility among many white southerners to teaching slaves to read and write, some slaves such as Frederick Douglass managed to do so. Indeed, according to antebellum U.S. censuses, free African Americans in the North often were more literate than Southern whites.[61] There is also interesting new work on Northern teachers who worked with recently freed African American slaves in the South during the Civil War.[62]

Much of the earlier revisionist debates about antebellum common schools focused on the motivations of education promoters as well as analyses of school attendance. Less attention was paid to what was actually happening in the classrooms. Some revisionists minimized the importance of antebellum teaching of intellectual or cognitive skills and emphasized schools' focus on inculcating discipline and work habits. Bowles and Gintis, for example, explained that "since its inception in the United States, the public-school system has been seen as a method of disciplining children in the interest of producing a properly subordinate adult population. Sometimes conscious and explicit, and at other times a natural emanation from the conditions of dominance and subordinancy prevalent in the economic sphere, the theme of social control pervades educational thought and policy."[63]

There is renewed interest today in examining the curriculum and textbooks used in the antebellum common schools and academies. As Carl Kaestle noted, many children attended rural single-classroom schools that included students of different ages and ability. As a result, teachers had to address a variety of different subjects and skill levels.[64] Children were expected to bring their own textbooks, usually readers, with little or no coordination on what books were to be used in the same class. Teaching initially focused on imparting the basic skills in reading, writing, and arithmetic. More advanced students received some education in subjects such as geography, history, grammar, and rhetoric, often relying upon their diverse, general textbook readers rather than more specialized school books.[65]

THE EMERGENCE OF PUBLIC HIGH SCHOOLS

Schooling beyond the elementary level was not widespread in early America. The Puritans, needing a well-educated ministry, established Harvard College in 1636. To prepare pupils for college, Boston and several other nearby communities supported Latin grammar schools to provide classical training. When a few more colleges were established in other colonies, additional Latin grammar schools were opened as well.[66]

In the eighteenth century, however, Latin schools faced competition from private academies that also prepared students for college admission. Some private academies, such as Phillips Exeter Academy, served mainly as college-preparatory schools; but most others also offered English training to attract non-college-bound students willing to pay the tuition.[67]

To help their non-college-going children in the increasingly commercial nineteenth-century economy, middle-class parents sought to provide them with an education beyond the common schools. Some parents also complained that talented and worthy youth from less affluent families and living in communities without a public Latin grammar school were unable to attend colleges because they could not afford a private academy.[68]

Boston created the first all-male English High School in 1821.[69] In 1827, the Commonwealth abandoned its requirement for larger towns to maintain Latin grammar schools, but mandated that communities with at least five hundred families create an upper-level public school that would teach American history, bookkeeping, geometry, surveying, and algebra (only cities with a population of five thousand or more had to continue teaching Latin and Greek). The law failed to clearly define what constituted a public high school or how it should be organized. Only a few Massachusetts communities immediately complied with the new law, but a large number of private academies were incorporated during the next decade, reflecting growing parental interest in providing additional education for their children.[70] William Reese, however, found that public high schools were opened in several cities in other states such as Baltimore, Cleveland, Cincinnati, Hartford, and Philadelphia.[71]

Earlier historians emphasized the lack of antebellum high schools as well as the fact that only a small minority of youth ever attended them. These scholars also assumed that most high school pupils came from affluent middle-class families rather than working-class homes. Therefore, it may seem surprising that one of the earliest and most influential revisionist studies of antebellum Massachusetts education, Katz's *The Irony of Early School Reform*, focused more on the debates about antebellum high schools rather than the attempts to reform the common schools.[72]

Katz analyzed the Beverly High School in the mid-nineteenth century—in large part because of the availability of a list of citizens who voted on whether to abolish that institution on the eve of the Civil War. Katz linked the societal divisions accompanying the shoemaker strike in Essex County with the opposition to the Beverly High School—especially among lower socio-economic workers: "Contrary to the myth that views public secondary education as the fulfillment of working-class aspirations, the Beverly vote revealed the social and financial leaders of the town, not the least affluent citizens, as the firmest supporters of the high school."[73] Indeed, according to Katz, the high school was imposed upon the Beverly workers by educational promoters.[74]

Katz's influential analysis of the abolition of the Beverly High School received considerable praise as well as criticism. In the early 1980s, Vinovskis—using more sophisticated multivariate statistical techniques and undertaking a more in-depth historical investigation of the community—re-analyzed the origins and temporary demise of the Beverly High School. Vinovskis agreed with Katz that the high school was imposed by local education proponents who used the threat of a substantial fine for failure to comply with Massachusetts education laws. Many Beverly citizens resented the creation of an expensive public high school when those monies might have been better spent improving common schools in the outlying areas. Rather than emphasizing the leadership of capitalists (especially the manufacturers) on behalf of a high school, however, Vinovskis discovered that it was the Beverly ministers who were its most numerous and active advocates on the local school committee. Nor was the town as bitterly divided along class lines on common school issues, as suggested by Katz's analysis. Most Beverly residents favored improving the common schools, though they were divided on the need for a high school.[75] Similarly, William Reese's study of nineteenth-century high schools found that local people often opposed these institutions even though they supported more common school funding.[76]

While revisionists challenged several traditional interpretations of schooling, they agreed with earlier scholars that few children entered antebellum high schools; and those who did attend largely came from middle-class families. Based upon a 10 percent random sample of Massachusetts towns with high schools in 1860, Katz concluded "that high schools were minority institutions probably attended mainly by middle-class children."[77] Katz also analyzed the occupations of fathers of Chelsea High School graduates (1858–1864) and fathers of students entering Somerville High School (1856–1861). He found that only one of the 141 high school students in those two Massachusetts communities (with the father's occupation known) came from a lower-class family.[78] Later, David Labaree reported that pupils in the Philadelphia Central High School in 1850 were less than one percent of the total number of common school students.[79]

The absence of working-class children in antebellum high schools did not seem surprising. Earlier studies of Massachusetts communities had already argued that few older immigrant and working-class youths attended schools. Stephen Thernstrom, in his seminal study of Newburyport social mobility, stated that Irish parents prematurely removed their children from school and sent them into the labor force. This strategy helped Irish families to purchase a home, but it deprived their children of the schooling necessary to compete for white-collar jobs. Thus, Thernstrom concluded that "opportunities for formal education past the age of ten or eleven, as a result, were effectively nil for working class children."[80]

Despite Thernstrom's widely cited assertion that few working-class children were enrolled in Newburyport schools past the ages of ten or eleven, an individual-level analysis of the 1860 census suggests otherwise. Almost every child in Newburyport living with one of their parents, regardless of the family's ethnic or occupational characteristics, attended school from the ages of six through twelve, including approximately 90 percent of children ages eleven or twelve from families with an unskilled male head of household. Moreover, a sizable proportion of those aged thirteen to nineteen also attended school; however, a multivariate analysis of those at the older ages showed that foreign-born children were less likely to attend than the native-born children.[81]

Even if the overall Newburyport school attendance was higher than suggested by earlier scholars, it does not necessarily mean that many of these older students attended a high school. However,

Newburyport, a medium-sized port city, had a well-developed school system that included three high schools—one for males, another for females, and a third coeducational institution. Based on lists of students enrolled in those institutions from 1857 to 1863 and the 1860 manuscript census, Vinovskis estimated that in 1860, 32 percent of all children aged eleven to sixteen enrolled in one of the high schools (and almost a third of them graduated). Children whose parents were native-born, in higher-level occupations, and more affluent were more apt to attend high school; but teenagers from more disadvantaged homes were not entirely excluded from high school. Almost one-sixth of the children of unskilled Newburyport fathers, for example, attended high school, and about one-fifth of them graduated. Children of foreign-born parents, however, were less likely to attend or complete high school than others.[82]

Was Newburyport an anomaly in terms of high school attendance in Massachusetts on the eve of the Civil War? Using an improved procedure for estimating the percentage of students who ever attended a high school, as well as community-level data for Essex County, Massachusetts, Vinovskis estimated that 15 percent of children attended a public high school. If one includes a rough estimate of those in private, secondary-level academies, an estimated 17 percent of Essex County children enrolled at some point in a secondary school.[83]

Can we reconcile the fact that Katz reported that 11.8 percent of Massachusetts children living in communities with a high school in 1860 attended, while Vinovskis' comparable figure for Essex County children residing in communities with high schools was 19.2 percent? If one recalculates the Massachusetts sample data, using the improved estimating methodology, Katz's high school attendance figure increases from 11.8 percent to 18.8 percent—very close to the Essex County figure. Thus, while only a small minority of antebellum children ever attended a high school nationally, in some areas such as Essex County and Massachusetts as a whole, the figures are considerably higher in 1860.[84]

We also need to acknowledge the diversity among communities with early high schools. Most antebellum communities only supported one high school; Newburyport with its three high schools was unusual. In large communities such as Boston or Philadelphia, this meant that only a small minority of youth ever attended because of the limited available seats. But in smaller and medium-sized communities, a much higher proportion of local youths could enroll. Indeed, in some smaller towns, students were actively recruited in order to justify the high cost of maintaining a high school. An analysis of high school attendance in 341 Massachusetts townships in 1875, for example, found that every one of the 55 towns with populations over 10,000 had a high school, as well as 73 of the 74 communities with 2,500 to 5,000 inhabitants. But even one-sixth of the smallest communities (under 1,250 people) had a high school. Overall, 58 percent of 1875 Massachusetts townships had high schools, compared to 31 percent fifteen years earlier.[85]

The total number of pupils in an urban high school usually was larger than in a rural township. As a result, one-third of the total number of students attending Massachusetts high schools in 1875 lived in urban areas (population 10,000 and up); nevertheless, one-fifth of all high school students resided in the smaller communities (population under 2,500); and almost half of all high school students came from communities with populations between 2,500 and 10,000.[86]

This complex picture of high school attendance invites more attention to how high schools were defined at that time. Due to the lack of a precise and standard definition of high schools in antebellum America, as William Reese has shown, high schools ranged from common schools that only added a couple of upper grades, to very elaborate and highly structured urban institutions. Most of the in-depth studies of antebellum high schools have focused on the larger urban high schools. As a result, we do not know as much about the so-called high schools in the smaller and medium-sized communities, or about the large regional differences in secondary education on the eve of the Civil War. Nor have we paid sufficient attention to private academies, which provided an alternative source of higher education for children who did not attend a high school.[87]

Did education foster social mobility in antebellum America? Revisionists questioned the advantages of education for working-class children and viewed schooling more as a way of training disciplined future workers; for middle-class students, however, revisionists saw common school education as yet another means to help parents transmit their already advantaged societal positions to their offspring.[88] During the nineteenth century, as schooling became more widespread, middle-class children continued attending school longer than their lower-class counterparts. Katz explained: "In this way, despite an overall rise in school attendance, the class differential in educational attainment has been preserved. Thus, despite the argument of early school promoters that education would reduce inequality, it is most likely that public school systems have reflected and reinforced existing social structures."[89]

On the other hand, Labaree, in his study of the Philadelphia Central High School, found that "students obtained admission to the school through a mixture of class background and academic ability. However, once admitted, they found themselves in a model meritocracy where academic performance was the only characteristic that determined who would receive the school's valuable diploma. Therefore, although middle-class students were still the primary beneficiaries of the high school, since they constituted the majority of those admitted, this class effect was mediated through a form of meritocracy that held all students to the same rigorous academic standard."[90]

Labaree's analysis suggested that once admitted to the Philadelphia Central High School, working-class students were able to compete with their middle-class counterparts in school. But Labaree did not trace their subsequent careers to assess the impact of high school attendance on children from different backgrounds. Reed Ueda, however, studied the inter-generational mobility of Somerville high school students in the second half of the nineteenth century and found that "the blue-collar son who was raised in the suburb and obtained the high school credential had powerful advantages over the average blue-collar son in Boston in obtaining white-collar employment. Blue-collar sons who went to high school in Somerville achieved a higher and faster rate of entry into the white-collar field than blue-collar sons in Boston of all levels of schooling."[91]

One of the most sophisticated studies of late nineteenth- and early twentieth-century social mobility concluded that high school attendance (and not necessarily graduation) helped most students' careers. Joel Perlmann's logit analysis of schooling and occupational mobility indicated that "the suspicion that secondary schooling did not help working-class boys, or immigrant working-class boys, who received it cannot be sustained. Education did not merely reflect the advantages of birth. Immigrant working-class boys who reached high school entered much more attractive occupations than others of similar social backgrounds."[92] Perlmann also found, however, that the while some African American students succeeded in high school, they did not fare as well in their careers due to the strong racial prejudices they encountered in the labor force.[93]

As more studies of antebellum high schools and private secondary-level academics are undertaken, more attention needs to be paid to the schooling of girls. Mary Kelley, for example, has just published a particularly interesting and challenging reinterpretation of private antebellum female academies and seminaries. She found that girls who attended these institutions became not only better educated, but also later assumed meaningful leadership roles in civil society. Kelley explained that the young women attending these female academies and seminaries between 1820 and 1840 were "being offered an education equal to the course of study at male colleges."[94] She noted that "comparisons of the numbers of women enrolled in female academies and seminaries show that, relative to male colleges, these schools were educating at least as many individuals in early nineteenth-century America. That approximately the same number of women and men were enrolled in institutions of higher learning is striking in its own right. It also provides the key to understanding why many women educated at these academies and seminaries pressed the boundaries that limited a woman's engagement with the world beyond her household."[95]

While most high school promoters stressed the benefits of those institutions for boys, Reese and other scholars have found that girls often enrolled in higher numbers and stayed longer.[96] Was the high school curricula for young women different from those of young men?[97] What impact did such an education have on the careers of young women, especially as better educated common school teachers? Did a high school education affect who these young women married or influence their adult experiences? Did girls who attended an urban high school differ from those who enrolled in a high school in a smaller community? Were girls in separate female high schools treated any differently than those who attended coeducational institutions? And did girls enrolled in private academies or seminaries have a similar curricula, or later pursue different careers than those attending public high schools?

CONCLUSION

The fields of colonial and antebellum education were substantially revitalized almost a half a century ago, though from rather dissimilar perspectives and generally by different scholars. On the one hand, Bernard Bailyn challenged historians to research colonial education from broader and more objective perspectives than his predecessors, such as Ellwood Cubberley. On the other hand, Michael Katz and the so-called revisionist scholars were simultaneously criticizing the previous positive portrayals of antebellum school reformers such as Horace Mann. Instead, the revisionists argued that antebellum common capitalists and their allies imposed schooling upon lower-class children in order to indoctrinate them to become more compliant workers.

Colonial education studies thrived as scholars explored diverse issues such as the extent and nature of male and female literacy, the growing availability of reading materials, and the impact of Puritanism on education. Ironically, as attempts were made to study education from a broader perspective, scholars neglected to document and analyze the roles of schools, churches, and local or colonial governments in fostering literacy throughout all of the colonies. European historians have been more successful in systematically investigating the varied relationships between early modern European education institutions and the development of mass literacy. American colonial historians would do well to pay more attention to these Europeans studies, both from a methodological and comparative perspective. This may also help to attract more scholars to the study of American colonial education, as well persuade their colleagues to pay closer attention to the impact of education on early American socio-economic, religious, cultural, and intellectual developments.

Sometimes the debates about antebellum education between the revisionists and their critics became unnecessarily acrimonious and personal in the 1960s and 1970s. But the challenging and useful questions raised attracted some of the best education analysts from a wide variety of scholarly disciplines. Studies of antebellum education also benefited from the relatively sophisticated use, for that time, of social science and qualitative analytical methods.

Today, most of the earlier rancor between the revisionists and their critics has disappeared as we develop a more complex and nuanced understanding of antebellum elementary and secondary education. More attention is now paid to important neglected education issues, such as regional and ethnic differences, as well as student and teacher classroom experiences. Much of the cultural and intellectual scholarship in this field continues to expand and improve, but few scholars today are still interested in, or capable of, using rigorous social science methods. This is unfortunate, as many of the key antebellum education topics awaiting further research will require both social science and qualitative analyses. Sadly, the field of antebellum education has lost much of its earlier interest and intellectual excitement, both from education historians as well as other scholars of social and cultural antebellum development.

One serious, but largely ignored impediment is that the separation of research between colonial and antebellum scholars has contributed to the large gaps in our knowledge of educational developments between 1760 and 1830. In his *Pillars of the Republic* a quarter of century ago, Carl Kaestle pointed out that many of the key transitional education and school developments occurred in the late eighteenth and early nineteenth centuries, a crucial time period often neglected by education historians. As scholars begin to reconsider these decades, both the study of colonial and antebellum education will be greatly expanded and enriched.

NOTES

1. See the discussion of this in Reese and Rury's introduction to this volume.
2. Bernard Bailyn, *Education in the Forming of American Society* (Chapel Hill: University of North Carolina Press, 1960), passim.
3. For example, see James Axtell, *The School Upon a Hill: Education and Society in Colonial New England* (New Haven: Yale University Press, 1974); David D. Hall, *Worlds of Wonder, Days of Judgment: Popular Religious Belief in Early New England* (New York: Alfred A. Knopf, 1989); E. Jennifer Monaghan, *Learning to Read and Write in Colonial America* (Amherst: University of Massachusetts Press, 2005).
4. In a study of the process by which America became modern, for example, Jon Butler has nothing to say about education, which as traditional theory suggests, was instrumental to the movement of traditional societies into modernity. Jon Butler, *Becoming America: The Revolution before 1776*, new ed. (Cambridge, MA: Harvard University Press, 2001). For a recent call to study the impact of social change on education as well as the influence of schools on society, see John L. Rury, *Education and Social Change: Themes in the History of American Schooling*, 2nd ed. (Mahwah, NJ: Erlbaum, 2005).

40   GERALD F. MORAN AND MARIS A. VINOVSKIS

5. Joel Perlmann, Silvana R. Siddali, and Keith Whitescarver, "Literacy, Schooling, and Teaching among New England Women, 1730–1820," *History of Education Quarterly* 37 (Spring 1997): 118–19.
6. R. A. Houston, *Literacy in Early Modern Europe*, 2nd ed. (London: Pearson Education, 2002), 125–39
7. Houston, *Literacy in Early Modern Europe*, 166.
8. Houston, *Literacy in Early Modern Europe*, 4.
9. Houston, *Literacy in Early Modern Europe*, 34.
10. Houston, *Literacy in Early Modern Europe*, 44.
11. Houston, *Literacy in Early Modern Europe*, 44–45.
13. Houston, *Literacy in Early Modern Europe*, 7.
14. Kenneth A. Lockridge, *Literacy in Colonial New England: An Enquiry into the Social Context of Literacy in the Early Modern West* (New York: W. W. Norton, 1974).
15. Lockridge, *Literacy in Colonial New England*, 4, 44.
16. Lockridge, *Literacy in Colonial New England*, 4.
17. Linda Auwers, "Reading the Marks of the Past," *Historical Methods* 13 (1980): 205. Auwers found that female mark-signature literacy rose from 27 percent, 41 percent, and 48 percent, respectively, for the birth cohorts 1650–69, 1670–89, and 1690–1709 and to 76 percent and 81 percent, respectively, for the birth cohorts 1710–29 and 1730–49.
18. Auwers, "Reading the Marks of the Past," 206. Because of state laws requiring the inclusion of girls in local schools, and because of growing support for female education from local churches admitting increasing proportions of women, Auwers argues that local Windsor schools lost whatever biases against female education that had presumably existed prior to the surge toward mass female literacy.
19. Auwers, "Reading the Marks of the Past," 212.
20. Perlmann, Siddali, and Whitescarver, "Literacy, Schooling, and Teaching among New England Women, 1730–1820," 120–21. Gilmore reported that while six out of ten women could sign in 1777–86 (similar to Main's data on rural Massachusetts in the 1760s), more than eight out of ten could do so in the period 1787–1806; Gloria L. Main, "An Inquiry into When and Why Women Learned to Write in Colonial New England," *Journal of Social History* 24 (Spring 1991): 581–86.
21. Joel Perlman and Dennis Shirley, "When Did New England Women Acquire Literacy?" *William and Mary Quarterly* 48 (1991): 50–67.
22. Perlmann, Siddali, and Whitescarver, "Literacy, Schooling, and Teaching among New England Women, 1730–1820," 139. For a thoughtful discussion of male and female schoolteachers, see Jo Anne Preston, "'He Lives as a Master': Seventeenth-Century Masculinity, Gendered Teaching, and Careers of New England School Masters," *History of Education Quarterly* 43 (Fall 2004): 350–71.
23. Kathryn Kish Sklar, "The Schooling of Girls and Changing Community Values in Massachusetts Towns, 1750–1820," *History of Education Quarterly* 24 (Winter 1991): 537.
24. Sklar, "The Schooling of Girls and Changing Community Values in Massachusetts Towns, 1750–1820," 537.
25. Main, "An Inquiry into When and Why Women Learned to Write in Colonial New England," 583.
26. Perlmann and Shirley, "When Did New England Women Acquire Literacy?" 64–65. In a recent essay, Daniel Howe has argued that "the American Enlightenment had made big plans for education in the young republic, but proved unable to carry them out effectively." Daniel Walker Howe, "Church, State, and Education in the Young American Republic," *Journal of the Early Republic* 22 (Spring 2002): 18.

27. Sheldon S. Cohen, *A History of Colonial Education, 1607–1776* (New York: John Wiley, 1974).

28. Elsie W. Clews, *Educational Legislation and Administration of the Colonial Governments* (New York: no publisher, 1899).

29. Gerald F. Moran and Maris A. Vinovskis, "Schooling, Mass Literacy, and Gender in America, 1630–1870," manuscript, 2007.

30. Farley Grubb, "Colonial Immigrant Literacy: An Economic Analysis of Pennsylvania, 1727–1775," *Explorations in Economic History* 24 (January 1987): 63–76; Farley Grubb, "Educational Choice in the Era before Free Public Schooling: Evidence from German Immigrant Children in Pennsylvania, 1771–1817," *Journal of Economic History* 52 (June 1992): 363–75.

31. Robert E. Gallman, "Changes in the Level of Literacy in a New Community of Early America," *Journal of Economic History* 48 (September 1988): 567–82.

32. David W. Galenson, "Literacy and the Social Origins of Some Early Americans," *Historical Journal* 22 (March 1979): 75–91. As Galenson has shown, the mark-signature literacy of 2,792 male indentured servants who emigrated from in the period 1718–1719 was "considerably higher than for the English population at large."

33. Monaghan, *Learning to Read and Write in Colonial America*, 242.

34. Mechal Sobel, *The World They Made Together: Black and White Values in Eighteenth-Century Virginia* (Princeton: Princeton University Press, 1987), chap. 14.

35. Janet D. Cornelius, *When I Can Read My Title Clear: Literacy, Slavery, and Religion in the Antebellum South* (Columbia: University of South Carolina Press, 1991).

36. Jill Lepore, *The Name of War: King Philip's War and the Origins of American Identity* (New York: Alfred A. Knopf, 1998).

37. Gaither, *American Educational History Revisited; and* Carl F. Kaestle, *Pillars of the Republic: Common Schools and American Society, 1780–1860* (New York: Hill and Wang, 1983), xi. Kaestle has provided a useful definition of common schools: "By 'common school' I mean an elementary school intended to serve all children in an area. An expensive independent school obviously, would not be a 'common school,' but neither would a charity school open only to the poor."

38. Ellwood P. Cubberley, *Public Education in the United States: A Study and Interpretation of American Educational History* (Boston: Houghton Mifflin, 1919).

39. Sidney L. Jackson, *America's Struggle for Free Schools: Social Tension and Education in New England and New York, 1827–42* (Washington, DC: American Council on Public Affairs, 1942); George H. Martin, *The Evolution of the Massachusetts Public School System* (New York: D. Appleton, 1902).

40. Ellwood P. Cubberley, *Rural Life and Education: A Study of the Rural–School Problem as a Phase of the Rural–Life Problem* (Boston: Houghton Mifflin, 1914); Thomas Woody, *A History of Women's Education in the United States*, 2 vols. (New York: Science Press, 1929).

41. Raymond B. Culver, *Horace Mann and Religion in the Massachusetts Public Schools* (New Haven, CT: Yale University Press, 1929); B. A. Hinsdale, *Horace Mann and the Common School Revival in the United States* (New York: C. Scribner's Sons, 1898); Louise Hall Tharp, *Until Victory: Horace Mann and Mary Peabody* (Boston: Little Brown, 1953); and E. I. F. Williams, *Horace Mann: Educational Statesman* (New York: Macmillan, 1937).

42. Bailyn, *Education in the Forming of American Society*; Cremin, *The Genius of American Education*; Cremin, *The Wonderful World of Ellwood Patterson Cubberley*.

43. For useful overviews of the history of education in the mid-1970s, see Sol Cohen, "The History of Education in the United States," in *Urban Education in the Nineteenth Century: Proceedings of the 1976 Annual Conference of the History of Education*

42   GERALD F. MORAN AND MARIS A. VINOVSKIS

Society of Great Britain, ed. D. A. Reeder (London: Taylor and Francis, 1977), 115–32; Lawrence A. Cremin, *Traditions of American Education* (New York: Basic Books, 1977); Michael B. Katz, "The Origins of Public Education: A Reassessment," *History of Education Quarterly* 16 (Winter1976): 381–407.

44. Michael B. Katz, *The Irony of Early School Reform: Educational Innovation in Mid-Nineteenth Century Massachusetts* (Cambridge, MA: Harvard University Press, 1968), 1.

45. Katz, *The Irony of Early School Reform*, 213–14.

46. Katz, *The Irony of Early School Reform*.

47. Samuel Bowles and Herbert Gintis, *Schooling in Capitalist America: Educational Reform and the Contradictions of Economic Life* (New York: Basic Books, 1976), 178–79.

48. Diane Ravitch, *The Revisionists Revised: A Critique of the Radical Attack on the Schools* (New York: Basic Books, 1978).

49. Walter Feinberg, Harvey Kantor, Michael Katz, and Paul Violas, *Revisionists Respond to Ravitch* (Washington, DC: National Academy of Education, 1980).

50. See Kaestle, *Pillars of the Republic*, 136–81 for a thoughtful, balanced discussion of the battles over antebellum common school reforms.

51. Carl F. Kaestle, *The Evolution of an Urban System: New York City, 1750–1850* (Cambridge, MA: Harvard University Press, 1973). Also see Stanley K. Schultz, *The Culture Factory: Bostom Public Schools, 1789–1860* (New York: Oxford University Press, 1973); and Selwyn K. Troen, *The Public and the Schools: Shaping the St. Louis System, 1838–1920* (Columbia, MO: University of Missouri Press, 1975).

52. Albert Fishlow, "The American Common School Revival: Fact or Fancy?" in *Industrialization in Two Systems: Essays in Honor of Alexander Gershenkron*, ed. Henry Rosovsky (New York: Wiley, 1966), 40–67. For a re-analysis of Fishlow's Massachusetts school data, see Maris A. Vinovskis, "Trends in Massachusetts Education, 1826–1860," *History of Education Quarterly* 12 (Winter 1976): 501–29.

53. Carl F. Kaestle and Maris A. Vinovskis, *Education and Social Change in Nineteenth-Century Massachusetts* (Cambridge, MA: Cambridge University Press, 1980); Lee Soltow and Edward W. Stevens, Jr., *The Rise of Literacy and the Common School in the United States: A Socio-Economic Analysis to 1870* (Chicago: University of Chicago Press, 1981).

54. Rury, *Education and Social Change*, 64. Rury stresses the long-term importance of nineteenth-century industrialization and urbanization on society and schools, but cautions us that "it would be wrong to assume, however, that there was a direct and immediate relationship between the rise of factories and the development of schools. In many instances the advent of industrialization meant that fewer children attended school in a given community."

55. Kaestle, *Pillars of the Republic*, 13–74; Maris A. Vinovskis, *Education, Society, and Economic Opportunity: A Historical Perspective on Persistent Issues* (New Haven: Yale University Press, 1995), 73–91. Although Bowles and Gintis at times acknowledged the differences in antebellum common schools, in general they focused heavily on Massachusetts. They justified this by pointing out that "the experience of Massachusetts was not perfectly replicated elsewhere, but we believe (and present some evidence) that the course of educational change in this state is not atypical of the rest of the country." Bowles and Gintis, *Schooling in Capitalist America*, 155.

56. On southern antebellum education, see Kaestle, *Pillars of the Republic*, 192–217; Bruce W. Eelman, "An Educated and Intelligent People Cannot be Enslaved: The Struggle for Common Schools in Antebellum Spartanburg, South Carolina," *History of Education Quarterly* 44 (Summer 2004): 250–70; Joseph W. Newman, "Antebellum

School Reform in the Port Cities of the Deep South," in *Southern Cities, Southern Schools: Public Education in the Urban South*, ed. David N. Plank and Rick Ginsberg (Westport, CT: Greenwood Press, 1990), 17–36; Kathyrn Walbert, "'Endeavor to Improve Yourself': The Education of White Women in the Antebellum South," in *Chartered Schools: Two Hundred Years of Independent Academies in the United States, 1727–1925*, ed. Nancy Beadie and Kim Tolley (New York: Routledge Falmer, 2002), 116–36; Jonathan Daniel Wells, "The Origins of the Southern Middle Class: Literature, Politics, and Economy, 1820–1880," PhD diss., University of Michigan, 1998, chap. 3.

57. Mary McDougall Gordon, "Patriots and Christians: A Reassessment of Nineteenth-Century School Reformers," *Journal of Social History* 11 (1978): 554–73; Kaestle, *Pillars of the Republic*; Ira Katznelson and Margaret Weir, *Schooling for All: Class, Race, and the Decline of the Democratic Ideal* (New York: Basic Books, 1985); David Tyack and Elisabeth Hansot, *Managers of Virtue: Public School Leadership in America, 1820–1980* (New York: Basic Books, 1982); Maris A. Vinovskis, *The Origins of Public High Schools: A Reexamination of the Beverly High School Controversy* (Madison, WI: University of Wisconsin Press, 1985); Julie M. Walsh, *The Intellectual Origins of Mass Parties and Mass Schools in the Jacksonian Period: Creating a Conformed Citizenry* (New York: Garland, 1998).

58. Kaestle, *Pillars of the Republic*; William J. Reese, *America's Public Schools: From the Common School to "No Child Left Behind"* (Baltimore: Johns Hopkins Press, 2005), 10–44; John Rury, *Education and Social Change*, 57–93; Maris A. Vinovskis, "Education and the Economic Transformation of Nineteenth-Century America," in *Age and Structural Lag*, ed. Matilda White Riley, Robert L. Kahn, and Anne Foner (New York: John Wiley, 1994), 171–96.

59. Nancy Beadie, "Female Students and Denominational Affiliation: Sources of Success and Variation among Nineteenth-Century Academics," *American Journal of Education* 107 (1999): 75–115; Nancy Beadie and Kim Tolley, eds., *Chartered Schools: Two Hundred Years of Independent Academies in the United States, 1727–1925* (New York: Routledge Falmer, 2002); Mary Kelley, *Learning to Stand and Speak: Women, Education, and Public Life in America's Republic* (Chapel Hill: University of North Carolina Press, 2006); Margaret A. Nash, *Women's Education in the United States, 1780–1840* (New York: Palgrave Macmillan, 2005).

60. Richard M. Bernard and Maris A. Vinovskis, "The Female School Teacher in Antebellum America," *Journal of Social History* 10 (Spring 1977): 332–45; Mary Carroll Johansen, "'Female Instruction and Improvement': Education for Women in Maryland, Virginia, and the District of Columbia, 1785–1835," PhD diss., College of William and Mary, 1996; Donald H. Parkerson and Jo Ann Parkerson, *Transitions in American Education: A Social History of Teaching* (New York: Routledge Falmer, 2001); Joel Perlmann and Robert A. Margo, *Women's Work? American School Teachers, 1650–1920* (Chicago: University of Chicago Press, 2001); John L. Rury, "Who Became Teachers? The Social Characteristics of Teachers in American History," in *American Teachers: Histories of a Profession at Work*, ed. Donald Warren (New York: Macmillan, 1989), 9–48; Maris A. Vinovskis and Richard M. Bernard, "Beyond Catherine Beecher: Female Education in the Antebellum Period," *Signs* 3 (Summer 1978): 856–69.

61. Cornelius, *When I Can Read My Title Clear*; Robert L. McCaul, *The Black Struggle for Public Schooling in Nineteenth-Century Illinois* (Carbondale, IL: Southern Illinois University Press, 1987); William S. McFeely, *Frederick Douglas* (New York: Norton, 1991); John L. Rury, "The New York African Free School, 1825–1835: Conflict Over Community Control," *Phylon* 45 (1983): 187–98; Thomas L. Webber, *Deep*

44 GERALD F. MORAN AND MARIS A. VINOVSKIS

Like the Rivers: Education in the Slave Quarter Community, 1831–1865 (New York: Norton, 1978).

62. Nina Silber, "A Compound of Wonderful Potency: Women Teachers of the North in the Civil War South," in The War Was You and Me: Civilians in the American Civil War, ed. Joan E. Cashin (Princeton: Princeton University Press, 2002), 35–59.

63. Bowles and Gintis, Schooling in Capitalist America, 37.

64. Kaestle, Pillars of the Republic, 13–59.

65. George H. Callcott, History in the United States, 1800–1860: Its Practice and Purpose (Baltimore: Johns Hopkins University Press, 1970); Douglas Alan Jones, "The Tradition of Didacticism in America's Early Reading Textbooks, 1780–1830," PhD diss., Michigan State University, 1990; Kelley, Learning to Stand and Speak; Cynthia Marie Koch, "The Virtuous Curriculum: Schoolbooks and American Culture, 1785–1830," PhD diss., University of Pennsylvania, 1991; Gerald F. Moran and Maris A. Vinovskis, "Schooling, Literacy, and Textbooks in the Early Republic," in A History of the Book in America, vol. 2 (forthcoming); Lisa Roberge Pichnarcik, "The Role of Books in Connecticut Women's Education in the New Republic: As Examined in Sarah Pierce's Litchfield Female Academy and James Morris' Coeducational Academy," master's thesis, Southern Connecticut State University, 1996; Edward W. Stevens, Jr., The Grammar of the Machine: Technical Literacy and Early Industrial Expansion in the United States (New Haven, CT: Yale University Press, 1995); Curtis Brent Wilken, "An Examination of American Reading Textbooks, 1785–1819, as an Expression of Eighteenth-Century Rhetorical Theory, and as a Precursor to Nineteenth-Century Writing Instruction," PhD diss., Ball State University, 2003.

66. Jurgen Herbst, The Once and Future School: Three Hundred and Fifty Years of American Secondary Education (New York: Routledge, 1996), 11–39.

67. Robert Middlekauff, Ancients and Axioms: Secondary Education in Eighteenth Century New England (New Haven: Yale University Press, 1963).

68. Nancy Beadie, "Internal Improvement: The Structure and Culture of Academy Expansion in New York State in the Antebellum Era, 1820–1860," in Chartered Schools: Two Hundred Years of Independent Academies in the United States, 1727–1925, ed. Nancy Beadie and Kim Tolley (New York: Routledge Falmer, 2002), 89–115.

69. It was initially called the Boston English Classical High School, but it was renamed three years later as the Boston English High School. On the origins of the Boston English High School and the difficulties of establishing a comparable school for girls, see William J. Reese, The Origins of the American High School (New Haven, CT: Yale University Press, 1995), 1–15.

70. Emit Duncan Grizzell, Origin and Development of the High School in New England before 1865 (New York: Macmillan, 1923); Alexander Inglis, The Rise of the High School in Massachusetts (New York: Teachers College Press, 1911).

71. Reese, The Origins of the American High School, 28–58.

72. Katz, The Irony of Early School Reform. When Katz analyzed public education systems in a later article, he noted that the date of the establishment of a high school was an approximate, but useful index of educational development. However, in his discussions of the public education and workers in that essay, he focused more on the common schools than the high schools. Katz, "The Origins of Public Education." For a useful discussion of the diverse analyses of common school reforms, see Kaestle, Pillars of the Republic, 104–35.

73. Katz, The Irony of Early School Reform, 19.

74. Katz, The Irony of Early School Reform, 85.

75. Vinovskis, *The Origins of Public High Schools*. Katz and Stevens were asked to write review essays of *The Origins of Public High Schools*, and Vinovskis agreed to respond. The three interesting and useful essays were published in *History of Education Quarterly*, 27 (Summer 1987): 241—58, and reprinted in Vinovskis, *Education, Society, and Economic Opportunity*, 126–41. Readers may also want to consult the 2001 introduction to the re-issue of Katz's book, Michael B. Katz, *The Irony of Early School Reform: Educational Innovation in Mid-Nineteenth Century Massachusetts* (New York: Teachers College Press, 2001), xiii–xxxix.

76. Reese, *The Origins of the American High School*, 59–79. On the debates over midwestern high schools, see David L. Angus, "Conflict, Class, and the Nineteenth-Century Public High School in the Cities of the Midwest, 1845–1900," *Curriculum Inquiry* 18 (1988): 7–31.

77. Katz, *The Irony of Early School Reform*, 39.

78. Katz, *The Irony of Early School Reform*, 270–71.

79. While Labaree does not try to calculate directly what percentage of children attended high school, the figure provided certainly suggests that the percentage would have been small. David F. Labaree, *The Making of an American High School: The Credentials Market and the Central High School of Philadelphia, 1838–1939* (New Haven: Yale University Press, 1988), 26–27.

80. Stephan Thernstrom, *Poverty and Progress: Social Mobility in a Nineteenth Century City* (Cambridge, MA: Harvard University Press, 1965), 23. For a sophisticated, statistical refutation of Thernstrom's assertions about the Irish, home ownership, and the school attendance of their children, see Steven Herscovici, "Ethnic Differences in School Attendance in Antebellum Massachusetts: Evidence from Newburyport, 1850–1860," *Social Science History* 18 (Winter 1994): 471–96.

81. Maris A. Vinovskis, "Patterns of High School Attendance in Newburyport, Massachusetts in 1860." Paper presented at the annual meeting of the American Historical Association, New York, December 28, 1985. Similar results for 1850 and 1860 were found by Herscovici, "Ethnic Differences in School Attendance in Antebellum Massachusetts."

82. Vinovskis, "Patterns of High School Attendance in Newburyport, Massachusetts in 1860"; Vinovskis, *Education, Society, and Economic Opportunity*, 142–56.

83. Vinovskis, *Education, Society, and Economic Opportunity*, 142–56. Also see David L. Angus, "A Note on the Occupational Backgrounds of Public High School Students Prior to 1940," *Journal of Midwest History of Education Society* 9 (1981): 158–84.

84. Vinovskis, *Education, Society, and Economic Opportunity*, 142–56.

85. Vinovskis, *Education, Society, and Economic Opportunity*, 157–70.

86. Vinovskis, *Education, Society, and Economic Opportunity*, 157–70. Another perspective is that an estimated one-sixth of all Massachusetts children in 1875 attended a high school at some point, and 18 percent of those living in a community with a high school enrolled. One out of ten Boston children ever attended high school as well as 14 percent of those in communities between 15,000 and 50,000 inhabitants. Only about one-eighth of children attended high schools in communities with fewer than 1,250 residents. The highest rates of high school attendance came in communities with a population between 1,250 and 10,000. And if we consider only those townships with a high school, children in areas with a population under 2,500 were much more apt to attend a high school than those in the urban areas.

87. Reese, *The Origins of the American High School*, passim.

88. Katz, "The Origins of Public Education," 397–98. Katz summarized that "the popular image of a continent wide open to talent simply cannot be sustained, though many men made modest gains that undoubtedly appeared critical to their lives. Few

laborers, that is, replicated the rags to riches version of success, but many managed eventually to buy a small house."

89. Katz, "The Origins of Public Education," 403. In a subsequent study of social mobility in Hamilton, Ontario in 1861, Katz and his colleagues downplayed the importance of education for social mobility even further: "School attendance itself, it is important to stress, did virtually nothing to promote occupational mobility. With other factors held constant, school attendance exerted no influence on the occupation of young men traced from one decade to another." Michael B. Katz, Michael J. Doucet, and Mark J. Stern, *The Social Organization of Early Industrial Capitalism* (Cambridge, MA: Harvard University Press, 1982), 275.

90. Labaree, *The Making of an American High School*, 37, 48–49. About a third of Central High School's students were from working-class homes—mainly those whose fathers were skilled workers.

91. Reed Ueda, *Avenues to Adulthood: The Origins of the High School and Social Mobility in an American Suburb* (Cambridge, MA: Cambridge University Press, 1987), 179.

92. Joel Perlmann, *Ethnic Differences: Schooling and Social Structure among the Irish, Italians, Jews, and Blacks in the American City, 1880–1935* (Cambridge, MA: Cambridge University Press, 1988), 38.

93. Perlmann, *Ethnic Differences*, 163–202.

94. Kelley, *Learning to Stand and Speak*, 29.

95. Kelley, *Learning to Stand and Speak*, 41.

96. Beadie and Tolley, *Chartered Schools*; Jurgen Herbst, *And Sadly Teach: Teacher Education and Professionalization in American Culture* (Madison: University of Wisconsin Press, 1989), 12–31, 57–86; Nash, *Women's Education in the United States, 1780–1840*; Christine A. Ogren, *The American State Normal School: "An Instrument of Great Good"* (New York: Palgrave Macmillan, 2005); Reese, *The Origins of American High School*.

97. Reese, *The Origins of the American High School*, 208–35; Kim Tolley, "Science for Ladies, Classics for Gentlemen: A Comparative Analysis of Scientific Subjects in the Curricula of Boys and Girls Secondary Schools in the United States, 1794–1850," *History of Education Quarterly* 36 (Spring 1996): 129–53.

CHAPTER 3

ALL EDUCATIONAL POLITICS ARE LOCAL:
NEW PERSPECTIVES ON BLACK
SCHOOLING IN THE POSTBELLUM SOUTH

Jacqueline Jones

By any measure, the founding of the Savannah Education Association in January 1865 represented a remarkable achievement for the African American community of that Georgia river port city. On New Year's Day, just a week after the occupation of Savannah by the army of General William Tecumseh Sherman, black leaders gathered in the First African Baptist Church and formed an organization to provide elementary schooling for the city's estimated 1,600 black children. An ecumenical mix of Baptist, Methodist, and Episcopalian ministers proceeded to constitute themselves as an executive board of the new Savannah Education Association (SEA). Two days later they met again, and in front of a large and expectant crowd, conducted a public examination of persons applying for teaching positions; by the end of the meeting, fifteen black Savannahians—ten women and five men—had received appointments as the first SEA instructors. The committee also drafted and approved a constitution that provided for a school board and a finance committee, and set fees for SEA membership on a monthly (25¢), annual ($3), or lifelong ($10) basis. In response to calls for community support, many in the audience came forward in what one observer called "a *grand rush* . . . Much like the charge of Union soldiers on a rebel battery," depositing a total of $730 in membership fees on a table in front of the committee. Many of these charter SEA members were men and women just shaking free from the chains of slavery—their precious dollars and pennies represented a striking commitment to the cause of education.[1]

On the morning of January 10, 1865, five hundred boisterous black children, shivering from the cold, assembled in the sanctuary of the First African Baptist Church. Spilling out into the street, they marched through Franklin Square and past the City Market, toward an imposing three-story brick structure on Market Square—their new school, the Old Bryan Slave Mart. For residents along the way, leaning out of windows and peering from doorways, this "army of colored children . . . seemed to excite feeling and interest, second only to that of Gen. Sherman's army," in the words of one amazed on-looker. In the old slave market, the pupils took their seats on wooden benches, their feet dangling above the floor. Surrounded by remnants of the old regime—handcuffs, whips, paddles, sales receipts for slaves—and positioned in front of the auctioneer's desk now occupied by their teacher, the girls and boys commenced their classes. Surely the African American community leaders who chore-ographed this grand school opening possessed a keen sense of the dramatic and the symbolic: black children would seize the streets, and they would transform the trappings of slavery into the means of their own liberation. Within weeks the SEA had in place five differ-ent schools, enrolling one thousand children.[2]

The Northern missionaries who entered Savannah hard on the heels of Sherman's army cast a suspicious eye on the SEA. Intending to establish a foothold of their own in the war torn South, agents of the American Missionary Association (AMA) in particular objected to the new black organization for several reasons. AMA agents, mostly ministers, had hoped to initiate their own effort on behalf of black schooling, and in the process win converts for the Congrega-tional denomination. They disapproved of the sight of black men and women who, though lacking formal teaching credentials, were usurping the places the AMA believed should be filled by white women graduates of Northern normal (teacher-training) schools. In February, one AMA agent, S. W. Magill, a Congregational minister from Connecticut, denounced the SEA as this "radically defective organization" run by black preachers "jealous of their honor and influence"—"leading men among the negroes [who] . . . have started on the principle of managing things themselves and just having their white friends do inferior work as assistants in carrying out their ideas and wishes." Together, representatives of the AMA and other relief organizations lamented the "rather peculiar feeling among the colored citizens here, in regard to the management of the schools . . . There is a jealousy of the white man in this matter. What they desire is assistance without control."[3]

The history of the SEA appears in several historical monographs exploring postbellum black schooling. Scholars (including the author of this essay) have interpreted the organization as a dramatic instance of the freedpeople's drive for autonomy, an attempt by former slaves to create and sustain their own schools, as well as their own churches and landed communities. Yet within a year and a half of its founding, the SEA had dissolved, unable to meet its financial obligations—deliberately deprived of needed funds by the AMA. The fact that the AMA eventually managed to subvert the SEA revealed the hostility of white Northerners of all kinds—military officials, agents of the federal Bureau of Refugees, Freedmen, and Abandoned Lands, no less than Northern missionaries—to thwart these various projects of self-determination on the part of black people all over the South. The brief history of the SEA thus encapsulates a now-familiar narrative that not only shapes more general scholarly accounts of black schooling in the postwar South, but also provides us with a stock cast of characters—the idealistic but naïve young white women teachers who came south after the war to teach the freedpeople; the heroic black Southern teachers, freedom-fighters all; the ineffectual Freedmen's Bureau officials, bent more on getting former slaves back into the cotton fields than into classrooms; and hostile, if not violent, neo-Confederates rabidly opposed to any initiatives related to black schooling.

Based on extensive research in archival sources, this narrative represents a considerable improvement over the original story offered by historians in the early twentieth century—a story that portrayed a vanquished white South, now besieged by Northern women busybodies and resentful former slaves, both groups bent on disrupting the traditional Southern social order that kept everyone in his or her proper "place." Yet the more recent interpretation is badly in need of updating and refining, now that we have learned more about the social and political dynamics of individual Southern communities. Local studies suggest several avenues of exploration that will help to complicate and enrich our understanding of the extent and meaning of black schooling during this turbulent period. Specifically, this essay focuses on four overlapping issues that we might collectively group under the topic of the politics of education: the significance of religious rivalries as a factor in black schooling in Southern cities; the complex forces shaping black parents' priorities for their children; the finances of freedpeople's education; and the link between education and labor in the rural South. To illustrate the fruitfulness of each of these areas of inquiry, I shall draw on my own recent study of Savannah and the Georgia low country between the 1850s and 1870s. Attention to local detail reveals the difficulty of generalizing about the motives of any of the major—or for that matter, minor—players in this Reconstruction drama. Throughout the South, schooling was deeply enmeshed in the fabric of local communities, and that fabric was often rent by conflicts that do not conform to what we might call the heroic interpretation of postbellum black schooling.[4]

The Georgia low country, including the river port of Savannah, presents a fascinating case study in the postbellum African American freedom struggle. For much of the nineteenth century, the city served as a commercial center for the transportation, processing, and distribution of rice, lumber, and cotton; these staples were brought by boat and train to Savannah's wharves from the coast and interior, milled and pressed, and then loaded onto ships bound for ports in the northeastern United States and Europe. Teamsters and dock-workers—so integral to this commercial economy—included large numbers of Irish immigrants, as well as black men, enslaved and free. (In 1860, the city's population included 22,000 people, with 8,400 blacks and almost 14,000 whites; among the latter group, one-third were immigrants.) On the countryside, rice slaves labored according to the exploitative task system, which forced them to complete a certain amount of work in the fields, and then forced them to devote time to growing food for their own families when their daily task was completed. Many enslaved men and women kept livestock and tended vegetables, which served as the source of marketable goods sold in Savannah. In contrast to slaves who toiled in the upland cotton economy, coastal slaves accumulated a variety of goods ranging from house wares to foodstuffs and horses, pigs, and chickens.[5]

In the city, blacks—regardless of legal status—imbued a spirit of enterprise; many slaves received permission from their masters to hire themselves out to work at their trades, and earned cash wages in the process. The SEA did not spring spontaneously from Savannah's liberation by the Union Army in late 1864; rather, the association had deep roots in the city's antebellum history. That history was marked by the influence of the independent black churches and their preachers, and by a patchwork of illegal, clandestine schools taught by black men and women. Not surprisingly, then, founders of the SEA were well-known educational and religious figures in the black community. Still, by early 1865, up and down the coast, the ligaments of the antebellum black community were undergoing a dramatic transformation. Thousands of refugees were crowding into Savannah, seeking safety from vengeful masters behind Union lines. Coastal slaves "refugeed" into the interior during the war were making their way back to their homes. Within this volatile situation, traditional patterns of religious leadership and family sustenance were changing rapidly.[6]

In Savannah, intense religious rivalries mirrored the turmoil engulfing every aspect of postbellum life. The black Methodists broke from the white parent body and allied with the African Methodist Episcopal Church. The city's three Baptist churches vied with each other for new members among the refugee population, and despaired over a loss of revenue in the face of the destruction of the city's commercial economy. It was within this context that established black leaders took alarm at the aggressive proselytizing among a series of AMA superintendents, most of whom saw elementary schools as a vehicle for converting large numbers of African Americans to Congregationalism. Though AMA emissaries appreciated what they considered black people's simple faith in God, these white men and women believed black worship styles to be excessively loud and demonstrative. Silenced under slavery, black congregants had found emotional release through church services, and responded accordingly to hymns and sermons. Both of these means of expression encoded the black freedom struggle, a struggle that would release blacks from bondage and usher in a "promised land" of freedom. Wedded to more staid and stiff worship services, and to a belief in the innate sinfulness of all humankind, the AMA made little headway in appealing to Savannah blacks or in prying them from their traditional religious leaders.[7]

In the summer of 1865, AMA officials began commissioning black preachers who would presumably serve as liaisons with the black community, and harvest souls for the Congregationalists in the vineyards of Savannah. Two black men took up the AMA standard— Hardy Mobley, a former Georgia slave who gained his freedom and moved north before the Civil War, and Robert Carter, the former slave of a famous Georgia politician, Howell Cobb. In June 1865, Mobley launched a vigorous campaign to bolster attendance at AMA schools at the expense of those sponsored by the AMA. However, he earned for himself only a "scurrilous attack" from the native black clergy. He soon abandoned Savannah for more promising fields of labor.[8]

Carter, attuned to the traditions of enslaved black congregants, fared little better. In 1869, C. W. Sharp, a newly arrived AMA superintendent, was claiming to his superiors, "There is a strong opposition, & a very bad state of things in Savannah, but if I can get a church built here, I believe a great field of usefulness will open." Sharp professed shock that all of the city's black preachers "are doing all they can, to raise suspicions, and throw doubt and discouragement our way." Seemingly oblivious to Mobley's failed mission, the AMA superintendent noted, "It is very trying. The very idea of our going ahead, seems to provoke them to . . . the exercise of their ingenuity in stirring up prejudice and awakening suspicions." Carter did his best to modify his own services and rituals in accordance with the expectations of potential new congregants. In an effort to appeal to Baptists, he agreed to depart from Congregationalist tradition and submerge converts, "or sprinkle [them], just as they like," as long as they joined his newly founded First Congregational Church of Savannah. Yet the church attracted only a handful of members.[9]

It is difficult to separate religious rivalries from funding issues during this period. A Catholic order of nuns—the Sisters of St. Joseph, trained in France for African missions—offered free classes for black children by the late 1860s. Impoverished parents found it virtually impossible to pay the one-dollar monthly tuition fee charged by the AMA, a policy designed not only to raise money but also, according to officials, to prevent the freedpeople from becoming overly dependent on white "charity." In 1869, one administrator turned out one-third of the pupils at the AMA's Beach Institute because they could not afford the monthly fee. At the same time, a number of small, private black schools were proliferating, as individual teachers tried to support themselves by charging less than the AMA. The talented black teacher Susie Baker King, a native of Savannah, made an effort to support herself and her son, but she found she could not compete with the AMA when it began to offer free night classes for adults. King was forced to close her school; she eventually found work as a domestic servant.[10]

By the early 1870s, the AMA knew that its missionary days were numbered. Composed of former Confederates, the Savannah city council had recently agreed to a takeover of the local Catholic schools. Parochial schools educated seven hundred Catholic children, and cost the local diocese $7,000 annually. Under a new arrangement (in the fall of 1869), the Savannah board of (white) education would accept financial responsibility for all of these Catholic schools and retain the current (Catholic) teachers. Those teachers were allowed to continue opening classroom exercise with scripture and prayer, and to use history books and other texts consistent with the Roman Catholic religion. The council's generosity reflected the political clout of the city's Catholic population, which included large numbers of Irish immigrants who worked on the railroads and docks. About this time, a number of prominent black preachers began to press the council to take control of Beach Institute, a large, new building that served as the showcase for AMA efforts in Savannah. The preachers wanted to rid the city of the AMA missionaries, and reassert control over the black schools. Council members agreed with the first goal, at least, and in 1872, began to offer free public education to a limited number of students (those who could squeeze into a tiny building that served as the first public black school). Although the AMA teachers maintained they would never consent to be turned over to the city "like furniture," within a couple of years the association had ceded its buildings to the board of education. For the next century, black children remained in segregated public schools, starved of taxpayer support and resources of all kinds.[11]

During the five or six years after the war, Savannah parents sought to exercise their own priorities—to the extent they were able—in the education of their children. In some cases, relatively well-to-do skilled tradesmen—some of them free before the war—scorned patronizing schools that attracted very poor pupils. Black preachers who opened their own schools appealed to the denominational loyalties of their congregants in seeking paying pupils. The AMA found that elderly white teachers could not compete with more energetic and popular black teachers who operated their own schools. Similarly, some parents preferred the independent schools that offered smaller classes, on the assumption that "they are getting more for their money for the reason that they have *more time spent on them*," and that the teachers were "*attending to them* [the children and parents] personally." In the fall of 1870, one AMA superintendent summed up the "hindrances [that] like armed men close in on every side": "The first families scoff at the idea of sending their children to school with refugees—Others think the advances in tuition a Yankee trick to get their money—others think that a bill before the Legislature will relieve them of the necessity of paying *anything*—others think that the colored people are competent to manage their own affairs in their own way."[12]

If the cultural arrogance and shortsightedness of the AMA superintendents is obvious here, the complicated factors affecting patterns of black religious leadership are less so. Preachers who had gained a substantial following before the war—such as the Reverend William C. Campbell of the First African Baptist Church—were used to tempering their language out of deference to the sensibilities of white people. After the war, these black men continued to deliver coded messages urging their congregants to embrace an earthly patience in expectation of a heavenly reward. However, they faced unexpected challenges from new leaders who now appeared on the scene, and proceeded to plunge into the fray of partisan politics. For the most part, these new leaders eschewed the tradition of dissembling in favor of a forthright call for black civil rights. In this sense, the antebellum customs of metaphor and rhetorical indirection were ill-suited for the times, which demanded direct engagement with the many crises besetting an emerging black citizenry.

Throughout this period, Campbell clashed with the Reverend James M. Simms—a former member of the First African Baptist Church—who escaped from Savannah during the war and fled to Boston, where he was ordained a Baptist preacher in 1864. Simms returned to Savannah in early 1865, and went on to become an SEA teacher, labor agent, member of the Georgia legislature, newspaper editor, local judge, and ubiquitous activist on the Savannah political scene. Campbell and others held that the clergy had no legitimate business running for office or otherwise agitating for political change. Yet there were indications that the conflict between the two men stemmed not so much from high principle as from bitter personality conflicts and professional rivalries.[13]

The Campbell-Simms feud was emblematic of larger clashes based on principles and politics, clashes that divided Savannah's postbellum black leaders. One revealing exchange took place between the elderly Savannah preacher, the Reverend Garrison Frazier, and the newcomer Aaron A. Bradley at a mass meeting in the Second African Baptist Church in December 1865. Frazier urged his listeners to comport themselves in an orderly manner—"You must not steal!"—so as not to anger resentful, vanquished white Southerners. Bradley was not a preacher; he had escaped from slavery some three decades before and found his way to the North, where he studied law. Now he had returned to his home state, armed with extensive knowledge of the Constitution and a defiant attitude that alarmed all white authorities—Northern and Southern, civil and military—in Savannah. At the December meeting, Bradley ridiculed Frazier's exhortations: How could black people "steal" if all low country wealth was the product of their own labor? Hard-working men, women, and children were merely appropriating what was their due. Bradley urged the crowd to *"resist, if necessary, at the point of a bayonet,"* agents of the Freedmen's Bureau or any other white men who attempted to deprive black people of their right to the land their tears and blood had watered over the generations. He went on to announce his intention to open his own school. Yet many of the older preachers were determined to retain their influence; in their eyes, the schools provided a dangerous base for upstarts of various kinds, whether the newcomer attorney Aaron A. Bradley or the young Savannah native Susie Baker King.[14]

In Savannah, educational politics remained a lethal mix of denominational rivalries, personality conflicts, money concerns, and divergent priorities among various groups of parents. In the countryside, matters were considerably less complicated. There, freedpeople had few if any options when it came to literacy instruction. The area between the Little Ogeechee and Big Ogeechee rivers is a case in point. Part of the expansive coastal Rice Kingdom—the Ogeechee, as it was called—was, before the war, home to large numbers of illiterate slaves who spoke a pidgin dialect called "Geechee" (a mixture of English and West African languages). With their strong African cultural traditions, and their isolation from white people, Ogeechee freedpeople early demonstrated a strong, and at times, militant impulse toward economic autonomy. They would seem to be unlikely supporters of two unmarried white women teachers from New England.

The Ogeechee, as it was called, was part of the original Sherman Reservation, a broad swath of the Georgia Sea Islands and coastline partitioned for prospective black landowners in the spring of 1865. The area was a particularly rich prize for black people who sought to claim parcels of these fertile rice lands, and then work them collectively. In the antebellum period, with slaves outnumbering whites by four to one, the area had produced 1.2 million pounds of rice annually. As early as March 1865, the Ogeechee Home Guard, a militia company, as well as other self-governing bodies, emerged on several plantations. By this time, dozens of former slaves had settled hundreds of acres on lands that would gradually revert to their former owners. In fact, in the summer and fall of 1865, planters launched largely successful legal challenges to the blacks' possessory titles in the district. Freedmen's Bureau agents sought to insure that, once they were either evicted or reduced to wage labor, the workers could claim the proceeds from crops grown that year. Yet in their "self-directed labor" and their tenacious hold on the land, Ogeechee blacks remained at odds with both planters and federal authorities. In 1866, with the encouragement of leaders of the Savannah Colored Union League—especially James M. Simms and others—Ogeechee workers formed their own Union Club, which held meetings on Grove Point Plantation.[15]

Families in the region struggled for some kind of control over their own productive energies. Rather than pay a toll of seven out of every one hundred bushels of rice to the Freedmen's Bureau—the fee for using a central rice threshing mill—the workers winnowed the rice by hand at home. Some resisted selling the rice altogether, "as they consider it the Lord's, given them to feed hungry mouths with." To the extent possible, they tried to negotiate for relatively favorable terms from a landowner. One wealthy planter complained to the bureau at the end of December that the laborers on his plantation "refused to contract with me *upon any terms*, for the ensuing year, and, also, refuse to leave the place, and threaten violent resistance to any effort to put them off the place." (Noted one worker more generally, "When a man has been burned in the fire once you cannot make him run in again.") Other points of negotiation included the share of the crop (one-half vs. one-third), and the presence of an overseer in the fields. Still, the people had only modest wants. A coachman for one wealthy family had always done "my duty religiously," he said, but his master would whip him nevertheless "*just because he could.*" Now the freedman wanted "a little strip of land for a garden where I can raise a few things jes to keep me along." According to one woman, "One meal a day and little bits of coffee and freedom is a great deal better than slavery."[16]

With its strong roots in Western Africa and its cultural isolation from whites in general, the Ogeechee would seem an unlikely place for a successful AMA school taught by two unmarried female New Englanders. Yet Ogeechee blacks welcomed the teachers, who were both energetic and resourceful. Esther W. Douglass of Brooksville, Vermont, and Frances Littlefield of Hallowell, Maine, preferred a rural outpost to the more comfortable circumstances of their Savannah coworkers. Before the war, the forty-one- year-old Douglass had wanted to evangelize among the Cherokee, but her mother would not allow her to go. By 1864, her mother was dead, leaving Douglass bound "by no family ties." In her application to the AMA, she professed her commitment to the freedpeople, but admitted that "a fondness for travel, and desire to see more of our country, come in also, as less worthy motives." Both she and Littlefield taught together in Virginia before receiving their Georgia assignment.[17]

Douglass and Littlefield set up housekeeping in the main residence of Grove Hill, which they dubbed "Spinster Hall," and opened a school they called "Ogeechee Institute." The landowner uttered a scornful prediction: "Those women in there think they are doing something great but those children only learn like parrots. They [the teachers] will soon come to the bottom of their brains." But the planter's disdain only served to energize the women.[18]

Indeed, they found a warm reception from people of all ages eager to learn. An elderly man told the teachers he hoped the community would use them well. Held in the parlor of their residence, the school drew children from seven neighboring plantations. On her first day, Douglass was startled by the sight of "120 dirty, half naked, perfectly wild, black children crowded on the floor." She could hardly understand them, for, "their language was to us, a confused jargon." With the exception of a few children raised as house slaves, the pupils had a hard time comprehending her as well. Supplied with only a few slates and some printed cards she tacked to the walls, Douglass nevertheless tried to uphold Northern-style standards of punctuality and classroom decorum. But some children had to walk as far as six miles to school on cold winter mornings, wading bare-foot through icy water, and they were frequently absent, "minding 'birds' or 'babies.'" The teachers also conducted a night school for adults; among their pupils was an elderly man who learned from the children in his home as well as from the teachers at school.[19]

On a typical day, the two women rose at 5:30 in the morning, breakfasted on fried hominy, and conducted a round of home visits before school began. They checked up on elderly and crippled shut-ins, reading a few Bible verses and praying, handing out quilts and coats sent from the North, dispensing a teaspoon of painkiller mixed with sugar and hot water for a stomach ache, camphor for cholera. With the help of Harriet Gaylord, another AMA teacher who regu-larly came out from Savannah, they sponsored an "industrial school," teaching women and girls how to cut patterns, baste, sew buttonholes, and make collars. The teachers also found time to write "begging" letters north to solicit cash, used clothing, and garden seeds. Douglass and Littlefield rejected the AMA's almost patholog-ical fear that distribution of clothing and rations would render the people permanently dependent on private charity and the federal government: "They need help as they are just starting in life and in short time they can support themselves as well as they could them-selves and 'massa' too."[20]

A description of the teachers preparing a Christmas dinner for the elderly on the plantation—a tasty repast of beef soup with rice, and crackers and ginger cakes—seems reminiscent of the antebellum plantation mistress, distributing food and good cheer to her slaves. The teachers were delighted with their reward—a song of thanks, praising Littlefield's dark eyes and Douglass' curly hair, followed by a little dance. Certainly, some of the dialogue recorded by Douglass between herself and freedmen and women evokes the deference rit-ual practiced under slavery. When she observed that a young man pounding rice, was "hard at work I see," he replied "O yes, Mis-sus . . . work brings the greenbacks now . . . There's nothing better than greenbacks." Still, Douglass and Littlefield offered a valuable service to the Ogeechee community. Soon after his teachers went north for the summer (in May 1866), a pupil named James Grant wrote a letter to AMA headquarters, asking that the two teachers return in the fall: "please to send them to the Grove Hill Plantation a gain for they have done so much good here and Have been so kind to the sick they we all feel that they are dear friends to us." He continued, "we all shall be very glad to behold their faces here with us again and also a minister for we will soon be alone to ourselves." He added, "I hope that these few lines may find you in health and that the Gospel is going on with Great Glory in the northern states."[21]

In the fall of 1866, Esther W. Douglass and Frances Littlefield returned to the Ogeechee District, where "the colored people welcomed us joyfully," alerted to the teachers' arrival because "Uncle Jack had seen us (though no word had been sent) in a vision." The welcoming party bore gifts of rice, peanuts, eggs, and potatoes. At the beginning of 1867, the people on Wild Horn plantation had signed a contract that allowed them to control all of the housing on the place, so that they could reserve a dwelling for the teachers—a stipulation the landowner, William Burroughs ("Mr. Rebel"), reluctantly agreed to. But as soon as the contract was signed, Burroughs evicted the two teachers. The people believed that Douglass and Littlefield represented a threat to the white man's authority; now literate, black workers would be able to read the labor contracts they signed. Burroughs was adamant, and the teachers made sorrowful preparations to leave.[22]

Nevertheless, their second season of teaching had inspired the two women to be more outspoken in their condemnation of the credit system coming into favor with white planters in the Ogeechee District. Wrote Douglass of the annual settlement in December 1866: "Injustice and oppression are on every side & I do not see how these people can ever have any of their own." She summarized the cruel bargain: "They raise the rice that fills the pocket books of those in charge of the plantations but are told that the rations have taken even more than their share of the crop and so the next year begins with a debt." Plantation merchants were charging two times as much as supplies cost in Savannah. Planters routinely drove off workers after the harvest without paying them. In one case, a landowner, sitting astride his horse and towering over a group of workers, flashed a roll of bills and promised to pay the men for loading a flatboat full of rice. Once the work was finished, he turned and galloped away, yelling back to them, "Now you may whistle for your pay." No wonder, then, that many people did not want to sign a contract, for "they do not get anything but a little corn & meat for their year's work." In March 1866, Douglass and Littlefield took their leave of the Ogeechee; they had accepted new teaching assignments on Daufuskie Island, South Carolina. In the winter of 1868, the Ogeechee would erupt into violence; freedpeople there spirited thousands of pounds of rice from the barns of white landowners, and declared their determination to resist annual labor contracts that left them little after a year of backbreaking labor. Over the next decade, the region came to symbolize—in the minds of whites—the restlessness of black field hands, and, ultimately, the slow death of the low country Rice Kingdom. But for a brief period of time, the Northern teachers had successfully integrated themselves into a community that, while alien to them on many levels, warmly embraced them and the school they started.[23]

Local conditions also shed light on why Northerners and Southerners, blacks and whites, and men and women became teachers of the freedpeople. Sorting out the motives of individuals is a difficult enterprise, but it is clear that a significant proportion desperately sought the modest, but predictable, paycheck that even the smallest school might provide. Although usually cast as young idealists, many of the Northern teachers sponsored by the AMA came from modest circumstances and worried about supporting themselves. Whether orphaned or not, they were dependent on friends and extended kin in the North; for them, the opportunity to earn $15 a month and to "do good" at the same time was an appealing prospect. At home in New England, their options for self-support were limited, especially if they did not teach school. Few found the alternative—domestic service—acceptable.[24]

Southern blacks and whites who had the ability and inclination to teach, grasped at the prospect of a monthly salary. Though $15 was a seemingly modest amount, it came in the form of cash, and it was regular to the extent that a school was adequately supported by the surrounding community, the AMA, the Freedmen's Bureau, or—beginning in the late 1860s—the Peabody fund. In Liberty County, southwest of Savannah, Harriet E. Newall was desperate to secure an AMA commission; the widow of a former missionary to India, and bereft of a stable source of income, she was living on a plantation and trying to support her small son. Her adult stepchildren were rabid rebels, but she was eager to provide elementary instruction for the freedpeople, young and old, who lived in her area. Certainly, she showed courage in teaching, but her letters also suggest that an AMA commission was her last best hope to provide for her son and remain on the plantation bequeathed her by her husband.[25]

Financial considerations were foremost in the minds of many native black teachers, including the Savannahians Susie Baker King and Hettie Sabbatie, single women who hoped to avoid working in the fields or in a white woman's kitchen. In some places, competition for a teacher's position could be fierce, since so much—at least in relative terms—was at stake. Savannah offered mainly seasonal work to black men, and black women had few choices for wage earning, with the exception of washing and ironing clothes and keeping house for whites. In rural areas many black people worked from year to year without seeing much cash at all. At the same time, rural schools were often fragile ventures, supported by parents who were field hands. At Elliot's Bluff, north of St. Mary's, Anthony Wilson tried to make a living by teaching school. A former sawmill employee, he had lost a hand in a workplace accident, which presumably made him unfit for manual work in the fields. But he found himself fending off persistent, even aggressive, threats from another young black man who wanted the job and the paycheck that went with it.[26]

To acknowledge the significance of school teaching as a source of income within an impoverished region is not to denigrate the motives of the teachers. At the same time, it would be inaccurate to posit extremes of idealism on the one hand, and money-grubbing on the other, furthering stereotypes of freedom-loving, native-born African American teachers contrasted with native-born whites who thought only in terms of dollars and not the welfare of their pupils. Moreover, in some areas, the educational activities of the Freedmen's Bureau—including hiring and boarding teachers, and building and renovating schools—could have a substantial impact on local economies. It is this larger picture that helps to illuminate the local dynamics of black schooling.

Throughout the low country, white landowners found themselves conflicted over the issue of black schooling. Soon after the end of the war, it became clear that even a primitive schoolhouse, and a teacher for it, was an asset for planters who wanted to avoid high rates of annual turnover among the families living on a plantation. However, some whites considered schooling a dangerous distraction for their hands, especially if the instructor was a suspect character bent on preaching ideas of equality among the freedpeople. Teaching at the Grant school—located in Glynn County at the intersection of five rice plantations north of Brunswick—James Snowden, a black native of New York, took pride in his pupils. He reported to his AMA sponsors, "After day school I have walked out in the rice fields, where most of my night scholars work; there you will see them with a spelling book in a bag around their necks, and every spare moment they have you will see them studying." At the same time, their "joy and gratitude" was tempered by the disapproval of their former masters, who "were and are very much against their having a school."[27]

Like freedpeople throughout the South, workers on the holdings of Charles C. Jones, Jr. (in Liberty County) and Frances Butler (in MacIntosh County) pressed their employers for schools, an issue that over time became just as contentious as negotiations over labor contracts. Charles Jones had remained firm in his decision to "*deny consent* to the establishment of a schoolhouse upon Arcadia land," on the assumption that "it would, in the present condition of things, be but an opening to complications, losses, etc. etc." In contrast, Butler had hired a white man, "a young country lad fresh from college," who quickly found that teaching black children their ABC's a task beneath him. He left; however, because the school was so popular among her workers, Butler hired another teacher, a young black divinity student from Philadelphia. Listening in on the children's recitations one day, she was startled "to hear them rattle off the names of countries, lengths of rivers, and heights of mountains, as well as complicated answers to arithmetic."[28]

Frances Butler and some of her neighbors believed that building schoolhouses on their land constituted sound labor relations, but insisted that the teacher refrain from political "agitation." One AMA teacher in the Brunswick area confirmed the worst fears of native whites when she set about organizing blacks into "Grant Clubs" (named for U.S. general and 1868 presidential candidate Ulysses S. Grant). Ellen E. Adlington had long been a boarder in the household of Virgil Hillyer, who was now representing Camden County in the state legislature. She marveled at the grapevine that appeared to connect all freedpeople in the area, and facilitated political organizing: "These people are like telegraph wires[;] what one knows all knows." But while Hillyer was away in Atlanta, his business partner, William T. Spencer, decided that the teacher had become too much of a liability to white Republicans in the area, and tried to cut off the funding for her school; in December 1868, a local bureau agent named Douglas Risley stepped in with some money from the Peabody Fund and saved it.[29]

Few Northern whites felt more keenly the dilemma posed by rural black schooling than Freedmen's Bureau agent Lieutenant Douglas L. Risley, formerly of the Forty-second Infantry Division. Risley was convinced that Brunswick needed a person who would not only teach the ABC's, but also "instruct the adults in their *rights, privileges, and duties.*" A nearby planter expressed willingness to establish a school on his property, but did not want anyone to "disturb the present pleasant relations existing between [himself] and his hands." Some planters stipulated that Northern white teachers not be employed at all, for when it came to educate the black child, "*they wish, they say, to educate him in their own way.*" At the same time, Risley feared that the irresponsible and provocative acts of a single teacher could wreck the educational enterprise in a region where black schooling was contingent on the favor of conservative white people. This dilemma played out in a feud between Risley and two AMA teachers in 1867 and 1868.[30]

Risley was looking for a discreet Southern black man, someone who could board with a black family, withstand the scorn of whites, and gain the respect of the freedpeople. However, in the late fall of 1867, he got much less—and much more—than he bargained for when the AMA assigned two white women to Brunswick. Twenty-nine-year-old Sophia Russell, daughter of a Maine carriage maker, was a veteran AMA worker; she had served in Hampton, Virginia, before coming to Georgia. Sarah H. Champney, thirty-nine, was also an experienced teacher (she had worked in Aberdeen, Mississippi), the daughter of a Massachusetts physician who had served as an army surgeon during the war. The two seemed to be adventurous enough; they preferred setting up housekeeping alone, for in Champney's words, "I have observed that w[h]ere a number live together there is more or less trouble." Unable to find their own lodgings, they at first consented to board with a Northern man, but soon found him to be "a man of grossly intemperate habits." Within a few days they were sharing a room in the main house of Hofwyl plantation, a situation so unpleasant that it caused Russell to exclaim, "it may be, after my warfare is ended, after I have left my cross, some soul from Hofwyl will meet me at God's right hand—if so Oh how rich, will be my reward!" Among other things, they hated the food, especially the griddlecakes laced with lard, "with just enough flour to keep them from running away."[31]

Before long, the two had opened a school in a pine barrens infested with fleas and mosquitoes. Teaching 104 pupils between them, they expressed shock for—as they commented in their November 1867 report—the children "are very dull and stupid; might as well have been raised in Africa for all they know of real civilization to say nothing of Christianity." The Brunswick blacks were, they opined, "the most ignorant & degraded of all this ignorant & degraded race . . . just like raw Africans." By January, the teachers had moved to a new place—a boarding house owned by a white woman. This situation had its own liabilities, for Champney and Russell deeply resented the fact that Northern money was being funneled to the unreconstructed Southern whites who furnished room and board to them.[32]

Champney expressed a deeper frustration: "I do not yet feel very much acquainted with the [black] people as we cannot have any of them coming to the house; nor are we allowed to visit them . . . " Risley had put them on notice that they should not see any blacks outside the classroom, even on missions of mercy to individual households; he insisted that such a reckless move would only inflame the feeling of the whites, and endanger the whole educational enterprise in Brunswick. Compounding the teachers' growing annoyance was the fact that, in late January, after an engagement that lasted just three days, the thirty-year-old Risley had married a woman twenty years his senior, Mary F. Moore. The teachers claimed that Moore, a "southern reb . . . *still perfectly despises the negro, & loves the lost cause.*" In other words, Risley was becoming "Southernized." Wrote Russell to AMA headquarters, "My cheeks burn with shame for the honor of the [U.S.] government." [33]

By early March, Champney and Russell had managed to go to housekeeping on their own, and they began clandestine after-dark visits. Attending a local black church service, the two found themselves "quite agreeably surprised at the good order & quietness of the congregation & good sense & intelligence of the preacher." The people were "as still & attentive as any white congregation." And in just a few short months, their school had improved immeasurably with pupils who now "would compare favorably with any white school of the same grade."[34]

Risley was not impressed. When he found out about their nighttime ramblings, he absolutely forbade them to do any more visiting—a policy he claimed was for their own protection. The teachers were indignant. Once again, they appealed to AMA headquarters, denouncing Risley as a lackey of the rebels, and requesting permission to continue their visits to the homes of black families. The response they received could hardly have been more disappointing: they were told that they were in danger of disgracing the association and giving "its enemies occasion to speak evil of it." That summer, Risley sought to have the last word with AMA headquarters, warning them against returning his two nemeses to Brunswick; he wrote Russell directly and accused her of being "a morally jealous and disappointed women, who naturally has an unfortunate propensity for gossip and strong appetite for scandal." Her propensities and appetites were evident in her attempt to "blacken" his character—and that of his bride—with rumors "that there was something wrong between Mrs. R & myself before we were married." Champney and Russell acquiesced and decided to accept an assignment in the west Georgia town of Cuthbert for the coming year; upon arrival, they received "there comes Hell" from the local white folks, a greeting the two women took in stride. Over the next months, they would battle the suspicion among whites that they had come as emissaries of the Republican Party, their trunks full of money for black ballot-box stuffers. At the same time, Brunswick's primitive conditions—both boarding and schooling—proved too rough for the two women newcomers, both veteran AMA teachers. They had moved to Staunton, Virginia, by January 1869. Risley was again appealing to the AMA for an adventurous man, one who could contend with what he considered the primitive culture of Brunswick blacks, and not complain about poor food, ill cooked.[35]

These examples from the history of postbellum schooling in Savannah and the Georgia low country suggest the difficulties of categorizing teachers and then making assumptions about their motives or resourcefulness, depending upon whether they were Northerners or Southerners, blacks or whites, men or women. Certainly a number of prominent actors in the Reconstruction drama defy the stereotypes. At times, it is even difficult to say with any assurance who was from the North or the South. Like Aaron A. Bradley, Harriet Jacobs was a fugitive slave who later returned to the South. Together with her daughter Louisa, in late 1865, Harriet Jacobs had secured a Savannah teaching commission from the New York Society of Friends. By now well known for her sensational account of abuse at the hands of her slave master, and her ordeal of hiding in an attic for seven years, Jacobs came to Savannah a seasoned worker among black refugees. The abolitionist-author of *Incidents in the Life of a Slave Girl* had spent the last two and one-half years of the war teaching and providing material relief for blacks in Alexandria, Virginia. There, she took pains to publicize to Northern benefactors the indifference, if not outright sadism of the camp superintendent, a white minister sponsored by the American Baptist Free Mission Society—a man "harsh and tyrannical to the people under his charge, but fawning and obsequious to those in authority."[36]

The Jacobs's work in Alexandria convinced them that the freedpeople would respond best to black teachers, men and women who understood that the impoverished former slaves were "quick, intelligent, and full of the spirit of freedom." Opening a school in January 1864, Louisa fully appreciated the potential of her pupils: "When I look at these bright little boys, I often wonder whether there is not some Frederick Douglass among them, destined to do honor to his race in the future." Given the chance, the freedpeople would provide for themselves, but this process would require time and patience.[37]

Though they paid for first-class tickets, Jacobs mother and daughter were forced to travel to Savannah in the steerage compartment, and, only after they strenuously protested did the captain of the steamship allow them to take their meals with the white passengers. When they disembarked at Savannah, they were greeted by a distressing sight—clustered on the docks, freedpeople who had arrived a few days earlier from the islands of Ossabaw and St. Catherine's, displaced by returning landowners. The miserable refugees "huddled around a few burning sticks, so ragged and filthy they scarce look like human beings." Nearby, black laborers went about their work—the men loading the ships and the women working for 25¢ a day, separating good from bad cotton fibers. Gesturing toward the cotton bales packed for export and detritus scattered around the bluffs, one woman exclaimed to the newcomers, "There you see our blood. Three hundred weight when the sun went down or three hundred lashes, sure!"[38]

The two teachers opened a school in the Savannah Freedmen's Hospital, which was run by Major Alexander T. Augusta, an African American surgeon whom they had met in Alexandria. Fearful of venturing outside the city, where a smallpox epidemic was raging, Harriet expressed admiration from afar for the work of James M. Simms. He was working as a Baptist missionary-labor agent among some four thousand freedpeople in the Ogeechee. Jacobs decried the typical rice plantation labor contract, which—she understood—was an ill-disguised effort by planters to strip black families of all semblance of self-sufficiency. Though approved by the bureau, the contracts were "very unjust. They [workers] are not allowed to have a boat or musket. They are not allowed to own a horse, cow, or pig. Many of them already own them, but must sell them if they remain on the plantations." At the end of the year, hands inevitably found themselves in debt to the landowner, beholden for rations and other supplies received on credit. Workers who left the plantation without permission of the owner—or who entertained friends or kin—faced cash fines of 50¢ a day, and in some cases, expulsion. Discouraged by the rising tide of violence engulfing Savannah in the summer of 1866, Harriet and Louisa Jacobs sailed for the North that July.[39]

Harriet Jacobs's commitment to the Georgia freedpeople raises intriguing questions: was she such an outspoken advocate for black rights because she herself had experienced the hardship and degradation of slavery in the South, or because she had lived for thirty years in the North and had absorbed the militancy of the abolitionist movement among free blacks in that part of the country? The answer is probably "yes" to both questions, but her experience highlights the difficulty of pigeonholing teachers as either Northern or Southern, among other labels. This example and others suggest the need to consider each teacher on his or her own terms. Further, we must realize that, in some cases, men and women were not particularly suited to this way of making a living; they simply lacked viable alternatives. The act of teaching was not necessarily and in all cases a conscious political act, as much as a way to provide for one's family in a time and place devastated by war.

In the context of postbellum educational politics in Georgia, the founding of the Savannah Education Association in January 1865 suggests not so much the culmination of the former slaves' desire for schooling, as an opening salvo in an effort that would become increasingly contentious and complicated in the coming years. That contention, and those complications, of course, are the stuff of history, challenging the stark black-white narrative of postbellum schooling, and subjecting the issue of education to the same kind of scrutiny accorded labor and politics. In exploring controversies surrounding black schooling within a local context, we give voice to the many parents, pupils, teachers, politicians, and planters who together made history during this fascinating chapter in the history of American education.

NOTES

1. "a *grand*": W. T. Richardson to M. E. Strieby, Savannah, 2 January 1865, Archives of the American Missionary Association, Amistad Research Center, Tulane University, New Orleans (hereinafter, AMA Archives). The AMA archives include thousands of letters arranged by date and state of origin; the collection is available on microfilm. See also, "General Sherman and the Freedmen," *Freedmen's Record* 1 (March 1865): 33; "Schools in Savannah, GA," *ibid* 1 (May 1865): 72; "Colored Free Schools," *National Freedman* 1 (July 1865): 197–8.

2. "army": W. T. Richardson to M. E. Strieby, Savannah, 2 January 1865, AMA Archives. For other accounts of the children's march and the new use of the Old Bryan Slave Mart, see Charles C. Coffin, *Four Years of Fighting with a Volume of Personal Observations* (Boston: Ticknor and Fields, 1866), 435; John T. Trowbridge, *The Desolate South, 1865–1866: A Picture of the Battlefields of the Desolated Confederacy* (Boston: Little, Brown, 1956), 271–2; "From Savannah," *Anglo-African*, February 11, 1865.

3. "radically": S. W. Magill to AMA Secretaries Savannah, 2 February 1865; "jealous . . . wishes": *ibid.*, 26 February 1865, AMA Archives; "rather": "Letter from W. C. Gannett," *Freedmen's Record* 1 (June 1865): 92.

4. Henry L. Swint, *The Northern Teacher in the South, 1862–1870* (Nashville: University of Tennessee Press, 1941). For examples of what we might call the heroic interpretation of the SEA and other local black efforts, see Heather A. Williams, *Self-Taught: African American Education in Slavery and Freedom* (Chapel Hill: University of North Carolina Press, 2005); Ronald E. Butchart, *Northern Schools, Southern Blacks, and Reconstruction: Freedmen's Education, 1862–1875* (Westport: Greenwood Press, 1980); Robert C. Morris, *Reading, 'Riting, and Reconstruction: The Education of Freedmen in the South, 1861–1870* (Chicago: University of Chicago Press, 1981); James D. Anderson, *The Education of Blacks in the South, 1860–1935* (Chapel Hill: University of North Carolina Press, 1981); Jacqueline Jones, *Soldiers of Light and Love: Northern Teachers and Georgia Blacks, 1865–1873* (Chapel Hill: University of North Carolina Press, 1980). My work in progress is titled "Savannah's Civil War" (forthcoming, Alfred A. Knopf).

5. Walter J. Fraser, Jr., *Savannah in the Old South* (Athens: University of Georgia Press, 2004); Dylan C. Penningroth, *The Claims of Kinfolk: African American Property and Community in the Nineteenth-Century South* (Chapel Hill: University of North Carolina Press, 2003).

6. Clarence Mohr, *On the Threshold of Freedom: Masters and Slaves in Civil War Georgia* (Athens: University of Georgia Press, 1986); Whittington B. Johnson, *Black Savannah, 1788–1864* (Fayetteville: University of Arkansas Press, 1996).

7. John Ernest, *Liberation Historiography: African-American Writers and the Challenge of History, 1794–1861* (Chapel Hill: University of North Carolina Press, 2004). See also, Charles Lawanga Hoskins, *Episcopalians in Savannah* (Savannah: St. Matthew's Episcopal Church, 1983); E. K. Love, *History of First African Baptist Church* (Savannah: Morning News, 1888); James M. Simms, *The First Colored Baptist Church of North America* (New York: Negro Universities Press, 1969; orig. pub. 1888).

8. "scurrilous": E. A. Cooley to Samuel Hunt, Savannah, 3 November 1865, AMA Archives; Joe M. Richardson, "'Labor is Rest to Me Here in the Lord's Vineyard': Hardy Mobley, Black Missionary During Reconstruction," *Southern Studies* (Spring 1983): 7–9; Joe M. Richardson, "The Failure of the American Missionary Association to Expand Congregationalism Among Southern Blacks," in *Church and Community*

Among Black Southerners, 1865–1900, ed. Donald Nieman (New York: Garland Press, 1994), 261–83.

9. "there . . . suspicions": C. W. Sharp to E. P. Smith 26 March 1869, AMA Archives; "or": Robert Carter to E. P. Smith, 15 June 1869, AMA Archives.

10. Susie King Taylor, *Reminiscences of My Life in Camp* (Athens: University of Georgia Press, 2006; orig. pub. 1906), 55.

11. "like": A. N. Niles to E. M. Cravath, 20 February 1872; Savannah City Council Minutes, October 28 and December 9, 1868 and August 18, 1869; Annual Report of Mayor John Screven, 1869–70, 19; Robert E. Perdue, "The Negro in Savannah," (PhD diss., University of Georgia, 1971, 140–45; Richard J. Shryock, ed., *Letters of Richard D. Arnold, M. D., 1808–1876* (Papers of the Trinity College Historical Society, Double Series, XVIII–XIX), 163.

12. "they": Abbie Johnson to E. M. Cravath, 28 November 1870, AMA Archives; "hindrances . . . way": A. N. Niles to E. M. Cravath, 28 October 1870, AMA Archives.

13. Love, *History of First African Baptist Church*; Simms, *The First Colored Baptist Church.*

14. "You": *Savannah Republican,* December 13, 1865; "resist": ibid., December 12, 1865; Joseph P. Reidy, "Aaron A. Bradley: Voice of Black Labor in the Georgia Lowcountry," in *Southern Black Leaders of the Reconstruction Era, ed.* Howard Rabinowitz (Urbana: University of Illinois Press, 1982), 281–5.

15. Karen B. Bell, "'The Ogeechee Troubles': Federal Land Restoration and the 'Lived Realities' of Temporary Proprietors, 1865–1868," *Georgia Historical Quarterly* 85 (Fall 2001): 377–78; Paul A. Cimbala, *Under the Guardianship of the Nation: The Freedmen's Bureau and the Reconstruction of Georgia, 1865–1870* (Athens: University of Georgia Press, 1997), 169–72, 338–40; Mart A. Stewart, *"What Nature Suffers to Groe": Life, Labor, and Landscape on the Georgia Coast, 1680–1920* (Athens: University of Georgia Press, 2002). For correspondence and reports related to the postbellum history of the Ogeechee, see for example, Charlotte Cheves to Davis Tillson, 7 October 1865; W. W. Deane to Davis Tillson, Savannah 15 October 1865; Report from the Reverend W. H. Tiffany, Grove Hall Plantation, 1 December 1865; A. P. Ketchum to Capt. Savannah, 15 June 1866, Unregistered Letters Received, Reel 25, Bureau of Refugees, Freedmen, and Abandoned Lands, Georgia Archives (M798), Record Group 105, National Archives (hereinafter BRFAL-GA [M798]), Washington, D. C. (available on microfilm).

16. Bell, "'The Ogeechee Troubles,'" 381; Cimbala, *Under the Guardianship of the Nation,* 171; "as": Hattie Gaylord to Samuel Hunt 1 February 1866, AMA Archives; "refused": Thomas Clay Arnold to H. F. Sickles, 27 December 1865, Unprocessed Freedmen's Bureau Collection, Georgia Historical Society, Savannah, GA; "my . . . slavery": Hattie Gaylord to Samuel Hunt, Grove Hill Plantation, 2 February 1866, AMA Archives; "when": Esther W. Douglass to Samuel Hunt, 1 February 1866, AMA Archives.

17. "by . . . motives": Esther W. Douglass to George Whipple, Brooksville, VT,1 September 1864, AMA Archives; Frances Littlefield to George Whipple, Boston, 12 December 1864, AMA Archives.

18. Esther W. Douglass, "Joy in Service—My Life in Service," No. 1, 11; *ibid.*, No. 2, "Grove Hill, Chatham County, Georgia, Oct., 1865," Esther W. Douglass Papers, Amistad Research Center, Tulane University, New Orleans, LA; Esther W. Douglas to Samuel Hunt, December 27, 1865, AMA Archives; Frances Littlefield to Samuel Hunt, Grove Hill Plantation, December 30, 1865, AMA Archives.

19. "120 . . . jargon": Esther W. Douglass, "Joy in Service—My Life Story" No. 2, "Grove Hill, Chatham County, Georgia, Oct. 1865," 13, Esther W. Douglass Papers;

"minding . . . good": Esther W. Douglass to Mrs. E. A. Adams, Grove Hill Plantation, Esther W. Douglass Papers.

20. Esther W. Douglass to Mrs. E. A. Adams, 14 April 1866, Esther W. Douglass Papers; Esther W. Douglass to Samuel Hunt, Grove Hill Plantation, 1 February 1866, AMA Archives; Esther W. Douglass to Samuel Hunt 3 April 1866, AMA Archives; "they": Frances Littlefield to Samuel Hunt Grove Hill Plantation, 1 February 1866, AMA Archives.

21. "hard . . . greenbacks": Esther W. Douglass to Samuel Hunt Grove Hill Plantation,1 February 1866, AMA Archives; "please . . . states": James Grant to George Whipple 30 May 1866, AMA Archives.

22. "the": Esther W. Douglass, "Joy in Service—My Life Story," 15; Bell, "'The Ogeechee Troubles,'" 382; Esther W. Douglass to E. A. Adams Savannah, 25 February 1867, Esther W. Douglass Papers; Esther W. Douglass to Theodore Adams Wild Horn Plantation, 16 February 1867, Esther W. Douglass Papers.

23. "Injustice . . . debt": Esther W. Douglass "[1866]," Dec. 24, 1866 entry; "they": Esther W. Douglass, "No. 2 [Wild Horn Plantation, 1867]; "now: Esther W. Douglass, "Joy in Service—My Life Story," 18, Esther W. Douglass Papers; Bell, "'The Ogeechee Troubles.'"

24. Jones, *Soldiers of Light and Love*.

25. See the series of Newall letters in the AMA Archives, Spring 1869; Ronald E. Butchart, "Perspectives on Gender, Race, Calling, and Commitment in Nineteenth-Century America: A Collective Biography of the Teachers of the Freedpeople, 1862–1875," *Vitae Scholastica* 13 (Spring 1994):15–32; Ronald E. Butchart and Melanie Pavich, "The Invisible Teachers: Southern Women in the Freedmen's Schools, 1861–1876," paper presented at the annual meeting of the Southern Historical Association, November 2005.

26. See for example, Whittington B. Johnson, "A Black Teacher and her School in Reconstruction Darien: The Correspondence of Hettie Sabbattie and J. Murray Hoag, 1868–1869," *Georgia Historical Quarterly* 75 (Spring 1991): 90–105.

27. "After . . . school": James R. Snowden, Glynn Co., 15 March 1869, AMA Archives.

28. "*deny*": Robert Manson Myers, ed., *The Children of Pride: A True Story of Georgia and the Civil War* (New Haven: Yale University Press, 1972), 1363; "a young": Leigh, *Ten Years*, 49; "to hear": *ibid.*, 85.

29. "These": E. E. Adlington to E. P. Smith, Berne, 1 August 1969, AMA Archives; William T. Spencer to Douglas G. Risley,23 February 1869, AMA Archives; William T. Spencer Berne, 3 December 1868, AMA Archives; E. E. Adlington to E. P. Smith, Berne, 9 December 1868, AMA Archives.

30. O. W. Dimick to E. P. Smith, Savannah, 23 December 1867, AMA Archives; "instruct . . . hands": Douglas R. Risley to C. C. Sibley Brunswick, 8 August 1867, Letters Received, Reel 14, BRFAL-GA [M798]; Report of Gilbert L. Eberhart to C. C. Sibley, Atlanta, July 30, 1867, Letters Received, Reel 17, BFRAL-GA [M798].

31. "I": Sarah Champney to E. P. Smith, N. Bridgewater, MA, 11 August 1868, AMA Archives; "a man": Douglas R. Risley to E. P. Smith, Brunswick, 28 November 1867; AMA Archives; "it": Sophia Russell to E. Smith, Brunswick, 13 November 1867, AMA Archives; "with": Sophia Russell to E. P. Smith, Brunswick, 12 November 1867, AMA Archives.

32. "are": Sophia Russell and Sarah H. Champney, Freedmen's Bureau Monthly Teachers' Report for November 1867, Hofwyl Plantation, Glynn County, Reel 21, Freedmen's Bureau Education Records [M799], Records of the Bureau of Refugees, Freedmen, and Abandoned Lands, Record Group 105, National Archives; "the":

Sophia Russell to E. P. Smith 12 November 1867, AMA Archives; Sophia Russell to
E. P. Smith, Brunswick, 31 January 1868, AMA Archives.
33. "I": Sarah Champney to E. P. Smith, Brunswick, 1 January 1868; AMA Archives;
"southern . . . government": Sophia Russell to E. P. Smith, 31 January 1868, AMA
Archives; Sarah Champney to E. P. Smith, Brunswick, 31 January 1868, AMA
Archives.
34. Sophia Russell to E. P. Smith, Brunswick, 1 March 1868, AMA Archives.
35. "its": Sophia Russell to George Whipple, Brunswick, 7 May 1868, AMA Archives;
"a . . . married": Douglas G. Risley to Sophia E. Russell Brunswick, 9 July 1868,
AMA Archives; Annie R. Wilkins and Margaret Burke to E. P. Smith Brunswick, 27
November 1868, AMA Archives.
36. "harsh": quoted in Jean Fagan Yellin, *Harriet Jacobs: A Life* (New York: Basic Books,
2004), 173–8.
37. "quick": *ibid.*, 167; "when": *ibid.*, 179; "Jacobs School," *Freedmen's Record* (March
1865): 41.
38. See also *Savannah Daily Herald*, November 11, 1865; "From Savannah," *Freedmen's
Record* (January 1866): 3–4; House Executive Docs., 39th Cong., 1st sess., no. 70
(serial set 1256), 52; Henry L. Swint, ed., *Dear Ones at Home: Letters from Contra-
band Camps* (Nashville: Vanderbilt University Press, 1966), 80.
39. "very": Yellin, *Harriet Jacobs*, 196.

CHAPTER 4

"AS IS THE TEACHER, SO IS THE
SCHOOL": FUTURE DIRECTIONS IN
THE HISTORIOGRAPHY OF AFRICAN
AMERICAN TEACHERS

Michael Fultz

From the 1880s through the 1920s, the adage, "As is the teacher,
so is the school," was commonplace in the rhetorical repertoire of
African American educators in the South. The essence of its meaning
lingered throughout the period of de jure segregation. Its expression
encompassed vital themes related to the need and demand for a
"sound professionalism" among the expanding number of African
American teachers in the region. Its significance flowed from a self-
evident logic implicitly understood, and fundamentally contested, by
both black and white southerners: the "fate of the race" depended
on its schools; the quality of those schools depended on the quality
of the teachers they had; and the quality of the teachers depended
upon their character, dedication, and professional training. Ambrose
Caliver, the first African American research specialist hired by the
U.S. Office of Education, reduced the issues to a single sentence,
"In the hands of the Negro teachers rests the destiny of the race."[1]

Given their direct connection to fundamental issues related to the
transmission of knowledge and values, social mobility, and "racial

uplift" (and to the ideologies in which these themes were embedded), it is more than a little surprising that black teachers have not been the subject of greater scrutiny within African American educational and social history. Indeed, in 1988 when Ronald Butchart published his insightful historiography, "'Outthinking and Out-flanking the Owners of the World': A Historiography of the African American Struggle for Education," his comment that, "teachers in black schools have been virtually ignored, except for the freedmen's teachers," was unfortunately all too accurate.[2] While the scope and depth of the research in this field has expanded significantly over the past two decades, many gaps remain.

REDISCOVERING BLACK TEACHERS IN THE EARLY EMANCIPATION PERIOD

In the early 1980s, three works, Jacqueline Jones's *Soldiers of Light and Love*, Robert Morris's *Reading, 'Riting, and Reconstruction*, and Butchart's *Northern Schools, Southern Blacks, and Reconstrction*, reinvigorated the historiography of the freedmen's teachers of the early emancipation period. Although their research questions varied, all three sought to achieve a richer, more nuanced, appraisal of the ideologies, motivations, and outcomes of the freedmen's aid societies of the period. All three highlighted an almost inescapable paternalism in the relationships between the northern societies and the southern African American populations whom they sought to "aid," and in a broader sense, all three accentuated the limits of educational reform as a strategy for Reconstruction. Both directly and indirectly, all three also prominently emphasized African American agency and initiative, and in doing so, rediscovered—in some sense—African American teachers.[3]

Black teachers have often been invisible in historical accounts of the freedmen's educational efforts, merged into the mass of impassioned African American educators and citizenry who, as W.E.B. DuBois has famously asserted, politically imposed the "Negro idea" of "public education for all" upon a reluctant South. Black teachers dot the landscape of DuBois's *Black Reconstruction* and Horace Mann Bond's *The Education of the Negro in the American Social Order*, but do not make a sufficient impression as to warrant extended discussion of their specific contributions.[4]

Thus, those not familiar with the semi-annual reports of John Alvord, the General Superintendent of Schools for the Freedmen's Bureau, were undoubtedly surprised to learn from Morris that, despite the voluminous literature that has been generated on the (implicitly white) "Yankee schoolmarms," by 1869 black teachers exceeded white teachers among those formally employed by the aid societies. Alvord had noted as early as his third report in 1867 that: "It is evident that *the freedmen are to have teachers of their own color*" (emphasis in original), and indeed over the next two years, the number of black teachers surged sharply, growing from around one-third of the teaching force in January 1867 (549 of 1,641) to slightly more than one-half (1,742 of 3,293) in July 1869. Although there was nothing "natural" about the Yankee-centric focus of the literature, or that it took over a century to begin to investigate the circumstances and implications of Alvord's observations, Morris was unquestionably correct that the narrowness of the literature had "distorted the group portrait."[5]

Explanations for this increase in the number of black teachers have not been fully explored. Perhaps, as Morris hypothesized, competition between black and white aid societies contributed to the rise, especially as these organizations encountered the stark realities of moving out from the urban areas and into the rural plains of the South, where resistance to white teachers among white southerners—and perhaps among black southerners as well—was notoriously vehement. Perhaps by the late 1860s the expenses and complications of providing housing accommodations for the white northern teachers had become significant hurdles.[6] Without question, however, African Americans themselves were a primary catalyst.

The opening chapter of James Anderson's *The Education of Blacks in the South, 1860–1935* is an outstanding treatment in this regard, centered on his arguments that "the foundation of the freedmen's educational movement was their self-reliance and deep-seated desire to control and sustain schools for themselves and their children"; that "the values of self-help and self-determination underlay the ex-slaves' educational movement"; and that "this underlying force represented the culmination of a process of social class formation that started years before the Civil War."[7]

To extend these arguments, however, it is important to add a perspective that the historiography of the early emancipation period has never sufficiently extrapolated: that, as part and parcel of the extraordinary quest for black schooling which characterized this era, as an integral component of this "self-sustaining" movement, the emergence of an African American teaching cadre was simultaneously realized. Most participants in the Yankee crusade did not understand the African American preference for black teachers, and often were left "puzzled and resentful." Yet, as Kathleen Berkeley has commented with discerning insight in her study of demands for African American teachers in Memphis in the early 1870s, "As blacks sought to ensure educational opportunities for themselves and their children against the backdrop of Reconstruction and New South politics, they found themselves trapped between the paternalism of their emancipators and the racism of their former masters . . . southern blacks expressed reluctance about pinning their hopes for a quality education on school systems staffed by northern whites and administered by southern whites."[8] It is unimaginable to believe that a recently freed ex-slave population would act otherwise.

One of Anderson's key contributions was to highlight the social significance of the indigenous "native" (Alvord's term) and Sabbath schools, whose whereabouts and numbers often went uncounted in the official reports. These African American schools were a "home effort," as Alvord put it, conducted with "no outside patronage from any quarter." Perhaps even more than the better known efforts then taking place in southern cities, "these were truly 'freedmen's schools'" as Betty Mansfield characterized them in her particularly comprehensive dissertation: "established by freedmen; taught by freedmen; supported by freedmen."[9]

These indigenous schools grew out of the clandestine educational activities carried out during slavery, multiplying exponentially during and after the war. Sometimes the teachers in these schools developed ties with the aid societies, as was the case with Mary Peake, whose legendary school in the shadows of Fortress Monroe, near Hampton, Virginia, was one of the first to be sponsored by a freedmen's aid society.[10] Most often, though, these educators have remained elusive, hidden from enumeration and history's spotlight. Especially in the cities, black teachers were often well educated; it is likely that in the countryside teachers with lesser qualifications prevailed, a policy problem which continued well into the twentieth century.[11] "It was a whole race trying to go to school," Booker T. Washington remarked, and in this context, it was sometimes the case, as one black woman commented, that while others with more formal schooling were very much needed, "I kin cair 'em a heap farther'n they is." The work of women like her as teachers among the freedmen, as active participants in a communal mass educational effort, has been underappreciated and underresearched.[12]

Also, as Butchart has begun to demonstrate in subsequent work, "the northern African American community was much more central to the process of southern intellectual emancipation than has ever been guessed." Butchart, whose work continues to energize this subfield, estimates that approximately 22 percent of northern teachers were black. Drawing on a sample from New York State, his analysis has revealed that although African Americans were a mere 1.2 percent of the population of New York State in 1865, they made up around 15 percent of the state's freedmen's teachers. Despite facing discrimination in their salaries and in other aspects of their assignments, black teachers "demonstrated a deeper commitment to the work than their white coworkers," averaging 3.7 years in the South compared with 2.5 years. His data also indicate that feminization was a less predominant factor among black recruits than among white recruits—nearly 44 percent of black teachers from New York were men, as compared with 23 percent of the white New York teachers.[13] A number of key questions remain, including the timing and pace of northern black volunteerism, the material and sociocultural foundations of diverse gender experiences in the North, the hiring and placement practices of the aid societies, and the avenues through which black demands were forwarded. Also, the influence of African American aid societies such as the African Civilization Society and African Methodist Episcopal Church are important factors to be considered.[14]

BLACK TEACHERS FOR BLACK SCHOOLS

One of the many reasons to investigate more fully the experiences of African American teachers during the early emancipation period is to achieve a better, more in-depth understanding of the evolution of— and the politics surrounding—the establishment of public school systems in the South from the early 1870s onward. From the start, for example, the call for African American teachers for African American schools was widespread, and continued unabated through the turn of the century. Rabinowitz was correct in his assertion that it was "blacks themselves who forced this change," and although I have disagreed with his interpretation that these political victories represented "half a loaf," my own work has confirmed his analysis of the basic contours of the struggle.[15]

Ideologically, it is important to recognize that the drive for black teachers broadened the foundation for arguments concerning black teachers' special pedagogical abilities and social responsibilities, perspectives that span the eighteenth, nineteenth, and twentieth centuries. Educational historians might more fully investigate the various components of these perspectives and their place within the "social vision" of black communities at various moments in time. This will likely include reframing and reconceptualizing accounts of intraracial debates over the integrated versus segregated schooling. For example, from the famous *Roberts* case in Boston in 1849–1850, through the various urban battles in the South in the late nineteenth century, and including sometimes angry confrontations in the North from the 1880s through the 1960s, it is clear that vocal segments of the black community favored (de facto) segregated schools as long as those schools provided employment for black teachers. This was as true for Atlanta in the 1870s as for Springfield, Ohio in the 1920s and Ocean Hill-Brownsville in the 1960s.[16]

African American petitions and protests frequently mentioned black teachers' ability to enter into "sympathy" with black schoolchildren, both in the classroom and through visits to their students' homes, the latter a central feature in an extensive list of extracurricular obligations for African American teachers throughout the pre-*Brown* period. Assertions of intraracial compassion, empathy, and understanding also included convictions regarding "culturally relevant" pedagogical techniques that African American teachers would employ, as well as expectations that, while serving as "the living textbook," black teachers would provide encouragement for their students and present them with "lofty ideals": "He [the student] is shown the highest and best in life and assured that he can obtain them as well as anybody else if he will only persist." Such statements on the racially conscious and moral roles of teachers also affirmed their own individual achievements in forging a dynamic "character," a social and educational ideal among African American educators during the late nineteenth and early twentieth centuries.[17]

Both before and especially after the turn of the century, as the Great Migration made its way North, African American proponents

of all-black schools in the North continued to advance these considerations, arguing that segregated schools: (1) increased employment
opportunities for black teachers; (2) fostered kinder, more sensitive,
and encouraging treatment for black schoolchildren; (3) promoted
the introduction and use of Negro history and other relevant curricular materials; (4) demonstrated African American competence and
abilities in teaching and administration; (5) promoted higher levels
of student scholarship, better attendance rates, and higher graduation rates from high schools in particular.[18]

Aside from V.P. Franklin's classic *Black Philadelphia*, Jack
Dougherty's *More than One Struggle*, Davison Douglas's *Jim Crow
Moves North*, and Adah Ward Randolph's study of the Champion
Avenue School in Columbus, the struggles of black teachers in the
North represent virtually untouched terrain for historians of African
American education.[19] In fact, concentration on issues of integrated
versus segregated schooling has undoubtedly skewed the literature.
A full range of research questions await attention: the formation and
expansion of black schooling in the North, both before and after the
World Wars I and II mass migrations; the daily grind of classroom
experiences, often under conditions of double and triple sessions;
training, hiring, placement policies (including reactions to the formal and informal practice of dual assignment lists); and comparative
questions regarding similarities and differences between African
American teachers in the North and South in terms of their classroom activities, their extracurricular obligations, and their community social roles.

This broader work might combine educational, social, and urban
history more fruitfully than in the past. To use a term popularized by
the late Lawrence Cremin, what configurations of formal and informal educative institutions did urban and rural African American
teachers and families develop to serve their short- and long-term
needs?[20] How and in what ways did black private schools—an under-
explored mainstay in black communities since the colonial period—
fit into the mix? One important issue is the broad array of social
services urban black communities negotiated in various cities across
the half-century between the 1870s and 1920s. This is a particularly
germane issue in the South, where, after all, education was only one of
a variety of public municipal services required by black neighborhoods
in these often nascent urban areas: transportation (including street
maintenance and lighting), water supplies, medical care, and libraries

were some of the others. "Where the sidewalk ends" was all-too-fre-
quently an apt description of the Negro section's border. Where did
education fit among the priorities at various points? How active were
black teachers and administrators in shaping these multiple
demands, and did their roles—like the struggles themselves—change
with the changing times? Black principals, for example, were instrumental in negotiating some of the first public black libraries in the
South prior to World War I, but after this period—and in rural
areas—their participation is not as clear. Can we *document* the contextualized social roles of urban and rural African American teachers
and principals in order to move beyond *assertions* of leadership and
middle-class status?[21]

FEMINIZATION OF AFRICAN AMERICAN TEACHING

Greater appreciation of context will also enhance investigations of
the feminization of African American teaching. It will also allow
researchers to utilize a range of theoretical perspectives—drawn
from economic, political, and gender studies, among others—to
investigate the full range of black women's educational opportunities, social philosophies, and activism.

Though U.S. census and state-generated data disagree, it is fair to
say that in 1890 the gender distribution of black teachers in the
South was close to 50/50, give or take a few percentage points
depending on source, with black women around ten percentage
points lower than white women's rates. By 1920, almost 83 percent
of black teachers were women, slightly exceeding rates among
whites. There have been virtually no comprehensive investigations
that have explored in depth the underlying factors fostering this feminization of teaching. These are the years in which segregation hardened in the South, when disfranchisement and the white southern
"educational awakening" sanctioned state- and county-level discrimination, trends which devastated the external contours of African
American education as funding gaps rose irrevocably and arrogantly.
Census data also indicate that the number of African American
teachers increased by around 58 percent across these years, but the
number of African American students attending school increased by
almost 75 percent. Thus, as segregation and discrimination in the
South institutionalized and hardened, the teaching cadre in black

schools feminized dramatically, with an ever larger percentage of
black women employed to teach in increasingly impoverished black
schools attended by growing numbers of black children.[22] It's not
surprising that in his 1911 Atlanta University study on *The Common
School and the Negro American*, DuBois commented that it was his
"firm belief that the Negro common schools are worse off than they
were twenty years ago . . . The wages for Negro teachers have been
lowered, and often poorer ones have been preferred to better
ones."[23]

We still await a gendered analysis of these patterns, delineating
the "opportunity costs" affecting the employment and retention of
black women and black men in teaching, as well as their interactions
with the noxious policy context they had to negotiate. Why did the
absolute number of African male teachers decline in fifteen states
from 1900 to1920, and by over 20 percent in nine of those states?
How intense was the competition among African American women
for jobs in urban and rural schools? In what ways did state and
county-level preferences and practices distort training, hiring, and
salary considerations? How did these issues interact and play out in a
context of undisguised oppression? Between 1900 and 1905, for
example, the state of North Carolina eliminated funding for four of
the state's seven black normal schools, even though all seven were
conducted in rented facilities, having "no buildings and equipment
which belonged to the State."[24]

While the need for in-depth studies of the years between the
1880s and the 1930s looms large, Ann Short Chirhart's *Torches of
Light: Georgia Teachers and the Coming of the Modern South* is an
excellent example of an astute, gendered investigation of black
teachers in the midyears of the twentieth century. In fact, Chirhart's
book is unique in that it also provides, through interviews and historical analysis, a thoughtful comparison of black and white teacher
experiences in Georgia, suggestively portraying the black family's
role in crafting pathways to teaching for young women. Chirhart's
well-taken argument is that "by balancing traditional moral, religious, and community values with gendered notions of professional
authority," teachers "not only carved out a professional status for
women but also established a cultural space in which their authority
prevailed—a classroom that stretched into the community." Another
welcome contribution is placing black teachers' religious commitments firmly into the analysis: "Teaching represented a respectable,

dignified career that offered women a chance to support themselves and their families as well as to serve the community and God."[25] Valinda Littlefield's comprehensive study has offered similar comments, recognizing that "from the outset, southern black education and the black church were interlocked" and that "the religious beliefs of many African American women teachers permitted them to put on the armor of God and enter the educational battlefield with a clear understanding of what was at stake both professionally and personally."[26] There are clearly many dimensions of African American family, gender, and religious experiences that might be explored to enhance our understanding of black schooling across both the South and the North.

These studies by Chirhart and Littlefield, along with the ongoing scholarship by Linda Perkins and newer work by Sonja Ramsey, complements the strong work which has been done over the past twenty years in black women's history.[27] In terms of this review, especially notable are those works which have directly and indirectly studied the extensive range of black women teachers' infrastructure-building activities, connecting school, church, health, and social welfare. Christine Woyshner's work on the National Congress of Colored Parents and Teachers, its state chapters, and its relation to the National Parent-Teacher Association, has also boosted the field.[28]

AFRICAN AMERICAN TEACHER TRAINING

Another mostly unexplored feature of African American education from the post-Civil War period onward is the development of teacher training facilities. References abound on the role of the Freedmen's Bureau and the American Missionary Association, in particular, in establishing several black colleges and universities—including Atlanta University, Fisk University, Howard University, among others—in the late 1860s. But what about the less prestigious, but more numerous, normal schools founded then and later?[29] We know very little about institutions like the Beulah Normal and Theological School, initiated by a black minister in Alexandria, Virginia in 1862, which provided upper-level instruction for approximately sixty prospective student-teachers by 1868. Did other normal schools follow the innovative pattern established by the American Freedmen's Union Commission's school in Lynchburg, where students first attended classes for a few months and then went out to open schools elsewhere in the county, teaching for few months before returning to their normal classes where the sequence would start again?[30] By 1871, there were approximately sixty-one normal schools producing black teachers, along with around eleven colleges and universities. Twenty years later, in 1892, the U.S. Commissioner of Education's report listed 25 colleges and universities for African Americans, with 791 collegiate-level students, and 38 normal schools, with 3,551 students enrolled. It is likely that more black women than black men took the normal school route for professional training than went to college (at least until the black public normals and land-grant schools grew into collegiate rank in the 1920s and 1930s), but studies and data are limited. As Glenda Gilmore suggests, it is also likely that the coeducational experiences of black students in normal schools and other higher education institutions had intra- and inter-racial social consequences, but for now that possibility is only an intriguing hypothesis.[31]

We need to know much more about this first generation of African American teachers in the post-slavery South, their education, and their experiences within their respective institutions and communities. We must also investigate the growth and development of teacher training for African Americans, and its evolution from normal schools to state teachers colleges and graduate-level institutions. The following brief case study offers a glimpse of some of the significant research topics that these types of investigations will yield.

THE EVOLUTION OF ALABAMA STATE UNIVERSITY

Alabama State University dates to 1866–67, with the founding of the Lincoln School in Marion, Perry County, Alabama by the local African American community. The next year, the American Missionary Association (AMA) leased the building and began to operate the institution as a normal school. A new building was constructed in 1869, with funding from both the Freedmen's Bureau and, as was typical of the times, a substantial contribution from "the colored people of Alabama."[32]

In December 1873, pushed by African American legislators, the Alabama legislature passed a bill to establish "A State Normal School and University for Colored Teachers and Students," contingent on the facility in Marion being donated to the state. The AMA complied. One section of the enabling legislation provided what seems to have been a fairly common administrative custom in the late nineteenth century: a tuition-free option if students pledged that they would teach in the state, in this case for at least two years after graduation.[33]

In 1887, Lincoln Normal was basically forced out of Marion after a confrontation with cadets at Howard College, a local white Baptist school. In the aftermath of this dispute, a fire—likely arson—burned down the Lincoln's main building. In response, an African American boycott forced a prominent white business into bankruptcy. The state legislature intervened, and the black school was ordered to relocate. African American communities in both Birmingham and Montgomery vied for the school—black residents in the former city raising approximately $3,000 and six acres of land, while those in the latter offered $5,000 and three acres. Montgomery won, over the territorial objections of Booker T. Washington. However, for the next year, until 1889 when the state legislature passed a bill funding a State Normal School for Colored Students at Montgomery, the institution operated on a shoestring budget as a private school in what was called the "Beulah Bottom" section of the city, utilizing as classrooms the Beulah Baptist Church, private homes, and storefronts. Over five hundred students were enrolled in 1888, mostly in the grammar grades, but four normal classes were maintained as well.[34]

An informative appraisal of the institution circa 1906 can be gleaned from a frank, but essentially positive, report written by a black school visitor, W. T. B. Williams, then in the early stages of an almost forty-year career assessing the state of African American education for organized philanthropy, particularly the Jeanes and Slater Funds. At this stage in its development, the State Normal School for Colored Students at Montgomery had 1,014 students—375 men and 639 women—divided into two departments: 553 in the elementary division, which subsumed nine years, and 461 in the five-year normal division.

Some 176 male and 241 female students lived with families in the city; there were no dormitories on campus as late as 1919. Williams noted that State Normal served to accommodate African American elementary and high school students for whom "the city makes no adequate provision." (Through 1919 Montgomery did not provide high school facilities for African American students, and those who sought advanced education past the seventh grade had to attend State Normal or leave the city.) He added that through a variety of overlapping social networks, the school "is becoming a sort of social centre for the colored people." The faculty numbered 27 teachers: 18 black women, 6 black men, 2 white women, and 1 white man. The academic program in the normal department was described by Williams as fairly strong, "favorable with the first two years work of the more ordinary Northern high schools. The graduates of this institution enjoy the reputation of passing very successfully and in large numbers the state teachers' examinations."[35]

State Normal's move toward full maturation began in the 1920s. As the 1920–21 school year opened, George Washington (G. W.) Trenholm—then State Supervisor of Teacher Training for Negroes (he had conducted summer institutes for black teachers in the state since 1911)—was tapped to serve as acting president, remaining in office until 1925 when he died unexpectedly. He was succeeded by his son, H. Councill Trenholm, who remained in office until 1963.[36]

The key institutional development of the 1920s was the normal school's advance toward collegiate status, first with the introduction of a two-year junior college program in the fall of 1920, and then the initiation of a four-year baccalaureate program in 1929, highlighted by the formal name change to the Montgomery State Teachers College. (The school's name would change twice more, to Alabama State College for Negroes in 1948, and to Alabama State University in 1967). Within an eleven-year period, the institution went from the equivalent of a high school with an enrollment of around 600, to a state-accredited teachers college that served more than 4,700 students during the 1929–30 academic year. In the following academic year, 1930–1931, eleven college-level curricula were offered to a collegiate enrollment of 430: three strictly two-year junior college programs (for elementary school teachers, home economics teachers, and an elementary teacher course for summer students); four four-year curricula (for those majoring in elementary education, languages, social studies, and science and mathematics); and four two-year sequences, paralleling the four four-year curricula, for those who had already attained a junior college degree and wanted a full baccalaureate. The next year, 1931–32, collegiate enrollment during the regular school year exceeded high school enrollment for the first time in school history: 455 to 424.[37]

Two other significant innovations were first implemented during the 1920–1925 period: (1) a move to the quarter system; and (2) enlarged in-service programs, both summer school and extension classes. Their effects were mutually reinforcing. The quarter system—dividing the school term into four quarters of sixty days each—represented an attempt to be responsive to the needs of practitioners and, in addition, to provide opportunities for those taking summer school courses to accrue college credits. Under this system, upper-level students could enroll at the beginning of any quarter, with three quarters of work constituting a full academic year. This system enabled in-service teachers who worked in schools with differing opening and closing dates a variety of possibilities to pursue advanced training.[38]

The summer quarter initiated in 1921 was not a new idea—six-week programs were offered at State Normal throughout the 1910s—but it offered a unique innovation: a ten-week, sixty-day session, with classes held on Saturday rather than just Monday through Friday. In 1924, a new twist was added: the sixty-day summer school was divided into two terms to accommodate varying schedules, and, in addition, a six-week term was offered for potential matriculates who could not enroll for the full spring quarter. Essentially, then, State Normal offered three summer terms.[39]

Summer school enrollment grew rapidly. As early as its second year in operation under the new format, summer school represented the largest single segment of State Normal's overall enrollment, and by the late 1920s this programmatic activity dominated all other forms of matriculation.[40] The summer program proved so popular that in 1927, branch summer schools were opened in Birmingham and Mobile, programs that continued through the early 1950s. (In 1937, a year-round branch junior college was initiated in Mobile; 262 individuals had graduated from this program by 1950.) About the only program that failed to catch on was a 1936 attempt to establish a branch summer school at Snow Hill Institute, a small private institution in the heart of Alabama's Black Belt region.[41]

Another program directed at in-service African American teachers that proved extremely popular was extension courses, which made their debut in four counties during the 1921–22 academic year. Like the summer school programs, these courses were promoted as providing "double credit," meaning that the State Department of Education would recognize them for renewal of teaching certificates, and also that those taking the courses could accumulate credits for eventual graduation at State Normal. By 1925–26, a Division of Extension Service had been organized as an administrative unit: that academic year enrollment in extension classes exceeded one thousand for the first time, with classes offered in at least twenty-five counties. The classes typically met on Saturdays, at two- or three-week intervals, for a total of eight meetings of three hours each. Overall, by the early 1930s, the Division of Extension Services was coordinating, in whole or in part: (a) a Teacher Placement Bureau; (b) required state teachers' institutes; (c) a statewide Oratorical Contest every April (conducted with the Elks organization); (d) an annual Older Boys Statewide Conference; (e) an annual girls Statewide Basketball Tournament in March; and, (f) a Statewide Academic Meet for high school seniors. In 1941–42, correspondence study was added to the Division's scope of activities.[42]

The 1940s saw significant movement toward university status. Although the school was criticized in state reports during this period as needing improvements in its physical plant, libraries, laboratory facilities, and faculty salaries, to a large degree these aspects of its operation were contingent upon state funding, which historically had been inadequate. On the other hand, the school's academic program remained strong, and even a 1944–45 state report noted that preparation of secondary teachers was "more extensive . . . than at the state teachers colleges for white students." Moreover, according to a 1944–45 survey, fully 86.9 percent of the faculty had either doctoral (15.5 percent) or master's (71.4 percent) degrees, substantially higher percentages than at Tuskegee (10.2 percent and 47.3 percent, respectively).[43]

In 1940, in reaction to the external push of the U.S. Supreme Court's 1939 *Gaines* decision requiring the provision of equal graduate and professional training, and with ongoing pressure from African American communities, a fifth-year graduate program was initiated. A class of eight students received their Masters in Education (M.Ed.) degrees in 1943; 354 students did so by 1952. The Alabama State Board of Education restricted the graduate school offerings to eight-week summer sessions until 1948, when the graduate program began regular-year operation. Also, in the 1940–41 academic year, Montgomery State Teachers College was authorized to permit students to work toward their B.S. degrees without taking all of the required courses in education or formally qualifying for their teacher's certificate. This move toward grafting a liberal arts college onto the ongoing professional and precollegiate programs was formalized in January 1947, when the school was authorized—as were the state's five white teachers colleges—to confer A.B. and B.S. degrees in addition to the B.S. degree in Education (elementary or secondary) and the M.Ed. degrees.[44]

INSTITUTIONAL AND SOCIAL RESEARCH AGENDA

Variations of this pattern of institutional development will likely be found in the evolution of many of the South's black public colleges and universities. By the mid-1930s, there were approximately thirty-five publicly supported teacher-training facilities for African Americans in the South—seventeen land-grant colleges, seven state teachers colleges, seven state normal schools, three municipal colleges, and one liberal arts college. Although Marybeth Gasman, Joy Ann Williamson, and Cally Waite, among others, have done impressive work in recent years investigating aspects of African American higher education, the histories of the black land-grants and other black public colleges still have not received the research attention they deserve.[45]

One significant theme that clearly stands out in the analysis of Alabama State is the role of summer schools in African American teacher training. One strand of this activity actually goes back to the early emancipation period, when black teachers began to implement summer instruction as a means of attracting students during their off-season from farming. Through the 1880s and 1890s, these summer schools also provided employment and training opportunities for countless numbers of African American normal and college students, as DuBois, for example, recounted in his bittersweet chapter in *The Souls of Black Folk*.[46]

Simultaneously, pushed by self-initiative and by demands for inclusion in the distribution of funds for Peabody Institutes, the tradition of summer schools for in-service training took root. After the turn of the century, and especially in the 1910s, this training vehicle was formally linked to changing state certification requirements, and was promoted from the mid-1910s through the latter part of the 1920s by grants from the General Education Board.[47] Summer sessions were such an important aspect of the in-service education of

African American teachers that, as data from Carter G. Woodson's *The Rural Negro* estimates, in 1929, 46.5 percent of black teachers in fourteen states attended seventy-seven summer schools; all states but one had a minimum of approximately one-third of its African American teachers attending summer schools, with highs of 97.7 percent of the employed teachers in Alabama and 69.2 percent of those in Mississippi.[48] (Often, especially through the 1930s, annual meetings of various black state teachers associations would be held in conjunction with the summer sessions.) How black teacher training programs of all types, public and private, adapted in multiple ways to the changing policy context of the 1910s and 1920s—a process which eventually fostered their transformation from normal schools to full collegiate rank by the late 1920s and early 1930s—is an important question to be addressed.[49] Note also that virtually from the onset of public schooling in the South, and especially during the decade and a half following the turn of the twentieth century, black teachers fought a running battle with various state governments, protesting attempts to implement second-class standards and lower licensing certificates that applied only to African American educators. African American representatives at a 1908 conference in Virginia summed up the prevailing sentiment:

> colored schools are at least as difficult to teach as any others, and accordingly as complete equipment should be required of the teachers of these schools as of those of any others . . . the proposed action will tend to lower the grade of colored teachers in general because it will give the legalized approval of the state to poor preparation on the part of colored teachers . . . it will not only not increase the number of good colored teachers in the state, but . . . will perpetuate the poorer ones and increase the number of inadequately prepared teachers . . . it will tend to stigmatize colored teachers in general as inferior.[50]

Another advantageous line of research might focus on collaborative and extension services, investigating direct and indirect connections between institutional practices and how teachers carried out their broad educative responsibilities. Frank appraisals should be included regarding the "home missionary" work black institutions and teachers felt was one of their foremost responsibilities. In Tennessee, the commitment began with faculty extension activities. As the school's 1914 catalog presented the extent of the endeavor:

All the members of the faculty are apostles of better living, and their field is the ninety-six counties of Tennessee. It is not merely institute extension, though they conduct a dozen summer institutes in as many counties. It is home missionary work. They go to the people and preach school libraries, individual drinking-cups, improvement of school grounds and school houses, home sanitation, village house cleaning, and the economic advantage of education . . . They visit meetings of the county courts, the bodies which appropriate the funds for the building of public schools, and plead their course. They are working everywhere to create social centers in the rural schools.[51]

However, we must realize that there is often a thin line between paternalism and "helping," between social class distain and "uplift." Varying assessments of how black educators negotiated these considerations, acknowledging both idealistic, well-intentioned efforts to assist (and "modernize") communities in need as well as the realities of entrenched community norms and practices, would certainly add a degree of perspective and sensitivity to a very complex set of issues.

It is also important to investigate the role black public colleges played in nurturing key components in the development and maturation of what was essentially a black educator-led infrastructure for racial uplift and social change. Again, to use Tennessee as an example, soon after Tennessee Agricultural and Industrial State Normal School (A&I) opened in 1912, participants at the summer school organized the State Rural Improvement Association. Before the session was over, twenty-one county-level organizations, with at least seven headed by women, were initiated throughout the state. The organization was subsequently headquartered at A&I and headed by the school's president, William Jasper Hale. In 1917–18, A&I became the headquarters for the state's African American Rosenwald Building Agent, and thus played a key role in rural school advancement statewide. In 1923, Hale and A&I hosted the founding meeting of the Tennessee State Association of Teachers in Colored Schools, the state's black teachers organization. A&I remained the headquarters for the teachers' association until its merger with its white counterpart in 1967. In addition, A&I was the headquarters for a number of African American county and home demonstration workers, as well as for the well-respected Jeanes supervising teachers of the state.[52]

It is worth noting that by 1906, when the Mississippi Association of Teachers in Colored Schools regrouped after an earlier effort had stalled, fifteen of what would eventually be twenty black state teachers associations had been founded, spanning all Deep South and border states. The umbrella organization, the National Association of Teachers in Colored Schools, was founded in 1904 (renamed the American Teachers Association in 1937). As with the normal schools and colleges, the clarion call of these organizations was to enhance the "professionalism" of African American teachers. (This was evident in how the founding and functioning of these groups was so often intertwined with issues of normal school training and the activities of the black land-grant colleges. For example, in Florida, Texas, Arkansas, Mississippi, and as noted in Tennessee, the president of the state's black land-grant school was instrumental in the founding of the state teachers group.) Held together by dedicated pioneers, before the 1920s membership in the teacher associations was low, often in the 100–500 range and seldom exceeding 1,000. By the 1940s and early 1950s, however, these black state teacher groups were mature and seasoned veterans of the southern school wars, led by full-time salaried executive directors. One highlight of their burgeoning activities was a series of unprecedented salary equalization lawsuits in the late 1930s and 1940s, initiated with the aid and support of the NAACP. (Note that, to date, the historiography lacks an incisive comprehensive analysis of these salary equalization drives.) In fact, by the post-WWII period, black state teacher associations were a central component within the infrastructure of innumerable local, county, state, and national black professional organizations that flowered during the era of de jure segregation. As Darlene Clark Hine has commented, "Without the parallel institutions that the black professional class created, successful challenges to white supremacy would not have been possible."[53]

A mere two decades after the *Brown* decision, however, by the early 1970s, the black state teacher organizations had been "absorbed." The institutional infrastructure supporting African American education in the South, and black teachers' social capital in general, had been critically altered. Nine of the final eleven black teacher groups merged with their white state counterparts during the heyday of integration following the passage of the 1964 Civil Rights Act. The national groups—the ATA and the NEA—merged in 1966. Terms were seldom equitable. As Rupert Picott, the longtime executive secretary of the Virginia Teachers Association remarked: "The absorption [of the VTA] . . . removed a power base for black teachers that has not been replaced. The resultant state teachers organization apparently has done very little to perpetuate the black teachers' association legend, memory and/or history." In several of the official histories of these groups, the tone verges on bitterness.[54]

African American teachers themselves were hard hit by the forces of "displacement" during the first two decades post-*Brown*. Throughout these years, black school staff at all levels—teachers, principals, coaches, counselors, band directors—were told formally and informally that their services were "no longer needed," as white communities throughout the South reacted first to the prospect and then to the reality of court-ordered desegregation. Black principals and other administrators were particularly decimated, as their schools were closed, and their students moved to the cities or were bused away. Between "displacement" on the one hand, and "one-way integration" on the other, memories and artifacts of the de jure segregated black schools in the South—including their teachers and staff—were eviscerated.[55]

Seeking particularly to "remember the good," Vanessa Siddle Walker's 1996 book *Their Highest Potential* struck a unique chord in African American educational historiography. "Bottom up" histories of the civil rights movement have altered traditional depictions of the dynamics of social change, and in *Their Highest Potential*, Siddle Walker combined history with ethnography to craft an intriguing new interpretation. Noting that constant characterizations of dilapidation and inferiority have "created a national memory that dominates most thinking about the segregated schooling of African American children," she cautions that "to remember segregated schools largely by recalling only their poor resources presents a historically incomplete picture." Some black communities did develop "good schools," characterized by high levels of academic achievement, positive affective experiences, and a family-like atmosphere—"cultural synchronization," as Jacqueline Irvine has called it—in which school staff and local communities worked together to enhance the children's educative experiences and provide personalized messages of hope in troubled times.[56] Though concerns over a historically unreliable nostalgia—"euphoric recall," as Siddle Walker calls it—hangs over some of this work, from a research and policy perspective, as Barbara Shircliffe has noted, "former students and teachers romanticized memories of their school experiences create an artful critique of the discriminatory aspects of the school desegregation process."[57] Along the way, new lines of investigation have been opened.

One of the important lines of inquiry which the "good segregated school" research has brought to the fore is the need to understand what are often sharply contrasting depictions of African American schools and African American teachers written before and after the 1980s. In many recent historical works, the treatment of black teachers is almost uniformly positive. However, prior to the 1980s, as noted earlier, black teachers *as women and as teachers*, were often absent from the historical narrative, and the manner in which they were depicted was sometimes quite negative. This difference in portrayal carries *meaning*, which has not been sufficiently probed.

If teachers were mainstays in the black middle class, why were their activities often disregarded? With the exception of the Jeanes supervisors, they certainly were not celebrated.[58] Teachers were fundamental to DuBois's conception of racial social development, yet neither his 1900 nor 1910 Atlanta University studies of black higher education recognized the variety of normal school services these institutions provided. Bond's *The Education of the Negro in the American Social Order* ignores black teachers as a subject of historical investigation, but rather presents various policy perspectives that attempt to explain the schools'—and their teachers'—inadequacies. Charles Johnson's depiction of black teachers in several of his sociological investigations is often unfavorable.[59]

One key issue is to unravel the perspectives of the generation of African American scholars and civil rights activists who came of age in the interwar years between the 1920s and the 1940s, whose academic work and policy prescriptions laid the foundation for the litigation and civil rights victories of the 1940s, 1950s, and 1960s. We lack thorough histories of the articulation and ascendancy of the integrationist ideology that characterized the 1920s—1960s period. How were various considerations weighed by various constituencies within African American—and white—communities at key points in this time period, especially during the quarter-century between the end of World War II and the early 1970s, when the push for integration was in its heyday? Was the interwar generation so concerned with overturning *Plessy* that whatever happened in its aftermath would be addressed as a matter of course? How are we to understand the incredible editorial comment by *Journal of Negro Education* editor, Charles H. Thompson, the dean of Howard University's School of Education, who wrote in 1951 that "the elimination of legally-enforced segregated schools should outweigh in importance the loss of teaching positions even by a majority of the 75,000 Negro teachers who might conceivably be affected"?[60] What did that generation really think would be the outcomes of integration, or how integration would affect not only African American teachers but also the black educational infrastructure which was then at its peak? Reassessing criticisms of black teachers and black college presidents in the 1950s and 1960s as timid, and even as "Uncle Toms," must be folded into the research agenda as well.[61]

What was not highlighted prior to the 1980s has become a central consideration in recent years. As Adam Fairclough has argued in his consistently impressive, thoughtful, and well balanced body of work, "the Civil Rights movement built upon earlier struggles and had its roots in preceding decades." As he elaborates, "in establishing schools and then struggling to raise standards, teachers helped to point the race in the direction of equality. "[62] In these ways, by developing a better understanding of how black teachers served as a "counterweight" to the public hostilities and crude conditions of segregation, how their classroom and community-building activities enhanced social capital in trying times, historians must begin to develop a more complex and holistic research agenda and a greater appreciation for the multiple struggles, limitations, and professional achievements of African American teachers.[63]

Notes

1. Ambrose Caliver, "Some Problems in the Education and Placement of Negro Teachers," Journal of Negro Education 4 (January 1935): 99. See also, "Radio Address of President Trenholm," *The Bulletin* 12 (December 1931): 11; Silas X. Floyd, "The Teacher and Leadership," *National Note-Book Quarterly* 2 (April 1920): 3; W. T. B. Williams, "Higher Education for Negroes in 1950," in *Proceedings, National Association of Collegiate Deans and Registrars in Negro Colleges, Eighth Annual Session*, March 7–10, 1934, 92.

2. Ronald Butchart, "'Outthinking and Outflanking the Owners of the World': A Historiography of the African American Struggle for Education," *History of Education Quarterly* 28 (Winter 1988): 361.

3. Jacqueline Jones, *Soldiers of Light and Love: Northern Teachers and Georgia Blacks, 1865–1873* (Chapel Hill: University of North Carolina Press, 1980); Robert C. Morris, *Reading, 'Riting, and Reconstruction: The Education of Freedmen in the South,*

1861–1870 (Chicago: University of Chicago Press, 1981); Ronald E. Butchart, *Northern Schools, Southern Blacks, and Reconstruction: Freedmen''s Education, 1862–1875* (Westport: Greenwood, 1980). See also, Sandra E. Small, "The Yankee Schoolmarm in Freedmen's Schools: An Analysis of Attitudes," *Journal of Southern History* 45 (August 1979): 381–402.

4. W. E. B. DuBois, *Black Reconstruction in America* (New York: Atheneum, 1970), 638; Horace Mann Bond, *The Education of the Negro in the American Social Order* (New York: Octagon Books, 1966).

5. J. W. Alvord, Third Semi-Annual Report on Schools for Freedmen, January 1, 1867 in *Freedmen''s Schools and Textbooks*, vol. 1, ed. Robert C. Morris (New York: AMS Press, 1980), 36. Robert C. Morris, "Educational Reconstruction," in *The Facts of Reconstruction: Essays in Honor of John Hope Franklin*, ed. Eric Anderson and Alfred A. Moss, Jr. (Baton Rouge: Louisiana State University Press, 1991), 150; emphasis in original.

6. Morris, "Educational Reconstruction," 150.

7. James D. Anderson, *The Education of Blacks in the South, 1860–1935* (Chapel Hill: University of North Carolina Press, 1988), 5.

8. Kathleen Berkeley, "The Politics of Black Education in Memphis, Tennessee, 1868–1881," in *Southern Cities, Southern Schools: Public Education in the Urban South*, ed. David Plank and Rick Ginsberg (New York: Greenwood Press, 1990), 202. Jones make a similar point concerning the situation in Georgia in the late 1860s, "black parents wanted their children to learn to read and write, but would not and could not yield them up irrevocably to the northern teachers." See her *Soldiers of Light and Love*, 111.

9. Anderson, *The Education of Blacks in the South*, 1–32; John Alvord, *Eighth Semi-Annual Report on Schools for Freedmen, July 1, 1869*, in *Freedmen's Schools and Textbooks*, vol. 1, ed. Robert C. Morris (New York: AMS Press, 1980), 5; Betty Mansfield, "That Fateful Class: Black Teachers of Virginia's Freedmen, 1861–1882" (PhD diss., Catholic University of America, Washington, DC, 1980), 194.

10. Morris, *Reading, 'Riting, and Reconstruction*, 1-2, 96-97; Dorothy Sterling (ed.), *We Are Your Sisters: Black Women in the Nineteenth Century* (New York: W.W. Norton & Co, 1984), 261-263; Kay Ann Taylor, "Mary S. Peake and Charlotte L. Forten: Black Teachers During the Civil War and Reconstruction," *Journal of Negro Education*, 74 (Spring 2005): 124-137.

11. See Ambrose Caliver, *Education of Negro Teachers(National Survey of the Education of Teachers, vol. 4)*. U.S. Office of Education, Bulletin #10. (Washington, DC: GPO, 1933); Bond, *The Education of the Negro in the American Social Order*, 263–83.

12. Anderson, *The Education of Blacks in the South*, 1–32; Butchart, *Northern Schools, Southern Blacks, and Reconstruction*, quoted on 127; Washington, quoted in DuBois, *Black Reconstruction in America*, 641–42; Heather Andrea Williams, *Self-Taught: African American Education in Slavery and Freedom* (Chapel Hill: University of North Carolina Press, 2005).

13. Ronald E. Butchart, "Recruits to the 'Army of Civilization': Gender, Race, Class, and the Freedmen's Teachers, 1862–1875," *Journal of Education* 172 (Fall 1990): 76–87, quote on 78; Butchart, "'We Best Can Instruct Our Own People': New York African Americans in the Freedmen's Schools, 1861–1875," in *African American Life in the Post-Emancipation South, 1861–1900*, ed. Donald G. Nieman (New York: Garland Press, 1994), 31–53. Similarly, Burchart's study of the freedmen's teachers from Oberlin College revealed that approximately one-quarter of the teachers who participated were African American, although black students had never been more than 5 percent of the school's enrollment body. See Burchart's "Mission Matters:

Mount Holyoke, Oberlin, and the Schooling of Southern Blacks, 1861–1917," *History of Education Quarterly* 42 (Spring, 2002): 1–17. See also Butchart, "Remapping Racial Boundaries: Teachers as Border Police and Boundary Transgressors in Post-Emancipation Black Education, USA, 1861–1876," *Paedgogica Historica* 43 (February 2007): 61–78. On the discrimination faced by African American teachers employed by the aid societies, see for example, Linda M. Perkins, "The Black Female American Missionary Association Teacher in the South, 1860–1870," in *Black Americans in North Carolina and the South*, ed. Jeffrey J. Crow and Floa J. Hatley (Chapel Hill: University of North Carolina Press, 1980), 123–36; Dorothy Sterling, ed., *We Are Your Sisters: Black Women in the Nineteenth Century* (New York: W. W. Norton, 1984), 261–63.

14. See, "An Appeal," *The Freedman"'s Torchlight* 1 (December 1866), unpaged in Morris, ed., *Freedmen"'s Schools and Textbooks*; Morris, *Reading, "Riting, and Reconstruction*, 13–14, 116–19; James M. McPherson, *The Negro"'s Civil War* (New York: Pantheon Books, 1975), 133–42; Patricia A. Young, "Roads to Travel: A Historical Look at the Freedmen's Torchlight—An African American Contribution to 19th Century Instructional Technologies," *Journal of Black Studies* 31 (May 2001): 671–98.

15. Howard N. Rabinowitz, "Half a Loaf: The Shift from White to Black Teachers in the Negro Schools of the Urban South, 1865–1890," *Journal of Southern History* 40 (November 1974): 565–94; Michael Fultz, "Charleston, 1919–1920: The Final Battle in the Emergence of the South's Urban African-American Teaching Corps," *Journal of Urban History* 27 (July 2001): 633–49.

16. Marcia Elaine Turner-Jones, "A Political Analysis of Black Educational History: Atlanta, 1865–1943" (PhD diss., University of Chicago, 1982); Philip Racine, "Atlanta's Schools: A History of the Public School System, 1869–1955" (PhD diss., Emory University, 1969); Davison Douglas, *Jim Crow Moves North: The Battle Over Northern School Segregation, 1865–1954* (Cambridge: Cambridge University Press, 2005); August Meier and Elliot Rudwick, "Negro Boycotts of Jim Crow Schools in the North, 1897–1925," *Integrated Education* 5 (August–September 1967): 57–68; "Springfield, Ohio, Defeats Segregation School Move," *Crisis* 26, no.1 (May 1923): 25; *Confrontation at Ocean Hill-Brownsville: The New York School Strikes of 1968* ed. Maurice Berube and Marilyn Gittell (New York: Preager, 1969).

17. See R. R. Wright, "The Possibilities of the Negro Teachers," *AME Church Review* 10 (April 1894): 459–70; Martha W. Owens, "The Development of Public Schools for Negroes in Richmond, Virginia, 1865–1900" (master's thesis, Virginia State College, 1947); Rabinowitz, "Half a Loaf," 579; Lena Jackson, "Intelligent Leadership," *The Broadcaster* 9 (January 1937): 29; T. W. Turner, "What the Colored Teachers of Baltimore are Doing for Their Race," *Colored American Magazine* 13 (July 1907): 36; Josephine Silone Yates, "Lincoln Institute: An Ideal Professional Training School for Negro Teachers," *Colored American Magazine* 12 (January 1907): 26–31. On the extracurricular obligations of African American teachers, see Michael Fultz, "African American Teachers in the South, 1890–1940: Powerlessness and the Ironies of Expectations and Protest," *History of Education Quarterly* 35 (Winter 1995): 401–22.

18. Roscoe Conkling Bruce, "The Stimulus of Negro Teaching," *Colored American Magazine* 17 (July 1909): 13–14.; Michael W. Homel, *Down from Equality: Black Chicagoans and the Public Schools, 1920–1941* (Urbana: University of Illinois Press, 1984); Judy Jolley Mohraz, *The Separate Problem* (Westport: Greenwood Press, 1979); W. E. B. DuBois, "Does the Negro Need Separate Schools?" *Journal of Negro Education* 4 (July 1935): 328–35.

19. Vincent P. Franklin, *The Education of Black Philadelphia: The Social and Educational History of a Minority Community, 1900–1950* (Philadelphia: University of Pennsylvania Press, 1979); Vincent P. Franklin, "The Persistence of School Segregation in the North: An Historical Perspective," *Journal of Ethnic Studies* 1 (Winter 1974): 51–68; Jack Dougherty, *More than One Struggle: The Evolution of Black School Reform in Milwaukee* (Chapel Hill: University of North Carolina Press, 2004); Jack Dougherty, "'That's When We Were Marching for Jobs': Black Teachers and the Early Civil Rights Movement," *History of Education Quarterly* 38 (Summer 1998): 121–41; Davison M. Douglas, *Jim Crow Moves North: The Battle over Northern School Segregation, 1865–1954* (Cambridge: Cambridge University Press, 2005); Adah Louise Ward Randolph, "A Historical Analysis of an Urban School: A Case Study of Northern De Facto Segregated School: Champion Avenue School, 1910–1996" (PhD diss., Ohio State University, 1996). See also, Linda Marie Perkins, *Fanny Jackson Coppin and the Institute for Colored Youth, 1865– 1902* (New York: Garland Press, 1987); Sulayman Clark, "The Educational Philosophy of Leslie Pinckney Hill: A Profile in Black Educational Leadership, 1904–1951," (PhD diss., Harvard University, 1984); John B. Reid, "A Career to Build, a People to Serve, an Purpose to Accomplish": Race, Class, Gender and Detroit's First Black Women Teachers, 1865–1916," in *""We Specialize in the Wholly Impossible"": A Reader in Black Women"'s History*, ed. Darlene Clark Hine, Wilma King, and Linda Reed (Brooklyn: Carlson, 1995), 303–20; August Meier and Elliot Rudwick, "Early Boycotts of Segregated Schools: The Alton, Illinois Case, 1897–1908," *Journal of Negro Education* 36 (August 1967): 394–402.

20. Lawrence Cremin, *Traditions of American Education* (New York: Basic Books, 1977).

21. On black principals' involvement in attempts to secure libraries in pre-World War I years, see Eliza Atkins Gleason, *The Southern Negro and the Public Library: A Study of the Government and Administration of Public Library Service to Negroes in the South* (Chicago: University of Chicago Press, 1941); Cheryl Knott Malone, "Accommodating Access: "Colored" Carnegie Libraries, 1905–1925" (PhD diss., University of Texas at Austin, 1996). On multiple urban needs, see Ronald H. Bayor, *Race and the Shaping of Twentieth-Century Atlanta* (Chapel Hill: University of North Carolina Press, 1996). See Adam Fairclough, "'Being in the Field of Education and Also Being Negro . . . Seems . . . Tragic': Black Teachers in the Jim Crow South," *Journal of American History* 87 (June 2000): 65-91, on the "deeply ambiguous" position of black teachers as community leaders; see also Fairclough, *Teaching Equality: Black Schools in the Age of Jim Crow* (Athens: University of Georgia Press, 2001).

22. See *Eleventh Census of the United States, 1890, Population, Part 2* (Washington, DC: GPO, 1895, cxviii and cxix; James Blodgett, "Report on Education in the United States at the Eleventh Census: 1890" (Washington, DC: GPO, 1893), Table 6, on page 51; David Blose, *Statistics of the Negro Race, 1927–28*, U.S. Office of Education, Pamphlet No. 14, (Washington, DC: GPO, December 1930); David Blose and Ambrose Caliver, *Statistics of the Education of Negroes, 1929–30 and 1931–32*, U.S. Office of Education, Bulletin 1935, #13 (Washington, DC: GPO, 1936). For literature which analyzes major aspects of the broader context, see Robert A. Margo, *Race and Schooling in the South, 1880–1950: An Economic History* (Chicago: University of Chicago Press, 1990); George M. Fredrickson, *The Black Image in the White Mind: The Debate on Afro-American Character and Destiny, 1817–1914* (New York: Harper & Row, 1971); Pamela Barnhouse Walters, David James, and Holly McCammon, "Citizenship and Public Schools: Accounting for Racial Inequality in Education in the

Pre- and Post-Disfranchisement South," *American Sociological Review* 62 (February 1997): 34–52.

23. See Carroll Miller and Howard Gregg, "The Teaching Staff," *Journal of Negro Education* 1 (July 1932): 196–223; W. E. B. DuBois, *The Common School and the Negro American* (1910; reprinted in *Atlanta University Publications* New York: Arno Press and the New York Times, 1969), 7. Although conditions improved in the 1920s and 1930s, overcrowded classrooms for black students and teachers persisted. According to Miller and Gregg, if the 47.1 average of pupils per teacher in the black elementary schools of the South in 1929–30 was to be reduced to the white pupil-teacher level of 34.3, an additional 17,190 African American teachers would need to be hired. If the national norm of approximately 30 pupils per teacher was to be reached, then more than 26,000 African American teachers were needed.

24. J.Y. Joyner to Wallace Buttrick, October 10, 1905, General Education Board Collection, Series 1, Subseries 1, box 113, folder 1022, Rockefeller Archive Center, Sleppy Hollow, NY; Charles Coon, "Announcement," July 1, 1905, quoted in W. A. Robinson, "A Campaign of Publicity for the State Negro Normal Schools of North Carolina" quote on page 1, Series General Correspondence, box 8, Folder Speeches, Reports, Misc, North Carolina State Archives, Raleigh, NC.

25. Ann Short Chirhart, *Torches of Light: Georgia Teachers and the Coming of the Modern South* (Athens: University of Georgia Press, 2005), 2–3; Ann Short Chirhart, "'Better for Us Than It Was For Her': African American Families, Communities, and Reform in Modern Georgia," *Journal of Family History* 28 (October 2003): 578–602.

26. Valinda Littlefield, "I Am Only One, But I Am One: Southern African-American Women Schoolteachers, 1884–1954" (PhD diss., University of Illinois, Urbana-Champaign, 2003), 127, 139. See also, Evelyn Brooks Higginbotham, *Righteous Discontent: The Women's Movement in the Black Baptist Church, 1880–1920* (Cambridge: Harvard University Press, 1993).

27. Linda M. Perkins, "The History of Blacks in Teaching: Growth and Decline Within the Profession," in *American Teachers: History of a Profession at Work*, ed. Donald Warren (New York: MacMillan, 1989); Linda M. Perkins, "The National Association of College Women: Vanguard of Black Women's Leadership and Education, 1923–1954," *Journal of Education* 172 (Spring 1990): 65–75; Linda M. Perkins, "Lucy Diggs Slowe: Champion of the Self-Determination of African-American Women in Higher Education," *Journal of Negro History* 81 (Spring 1996): 89–105; Sonya Yvette Ramsey, *Reading, Writing, and Segregation: A Century of Black Women Teachers in Nashville* (Urbana: University of Illinois Press, 2007).

28. See Christine Woyshner, "Black Parent-Teacher Associations and the Origins of the National Congress of Colored Parents and Teachers, 1896-1926," paper presented at the annual meeting of the American Educational Research Association (AERA), April 2000; Christine Woyshner, "The PTA and Desegregation: The Case of Alabama, 1954-1970," paper presented at AERA, April 2007. See also, Jacqueline Anne Rouse, *Lugenia Burns Hope, Black Southern Reformer* (Athens: University of Georgia Press, 1989); Darlene Clark Hine, Elsa Barkley Brown, Rosalyn Terborg-Penn, eds., *Black Women in America: An Historical Encyclopedia* (Brooklyn, NY: Carlson, 1993); Darlene Clark Hine, "'We Specialize in the Wholly Impossible': The Philanthropic Work of Black Women," in *Lady Bountiful Revisited: Women, Philanthropy, and Power*, ed. Kathleen D. McCarthy (New Brunswick: Rutgers University Press, 1990), 70–93. Stephanie J. Shaw, *What a Woman Ought to Be and to Do: Black Professional Women Workers During the Jim Crow Era* (Chicago: University of Chicago Press, 1996); Dorothy C. Salem, *To Better Our World: Black Women in Organized Reform, 1890–1920* (Brooklyn, NY: Carlson, 1990); Cynthia Neverdon-Morton, *Afro-American*

Women of the South and the Advancement of the Race, 1895–1925 (Knoxville: University of Tennessee Press, 1989); Sharon Harley, "Nannie Helen Burroughs: The Black Goddess of Liberty," *Journal of Negro History* 81 (Spring1996): 62–72; Higginbotham, *Righteous Discontent.*

29. James M. McPherson, *The Struggle for Equality* (Princeton: Princeton University Press, 1977), 402–3; Butchart, *Northern Schools, Southern Blacks, and Reconstruction,* 100–103; W. E. B. DuBois, ed., *The College-Bred Negro* (1900; reprinted in *Atlanta University Publications* New York: Arno Press and the New York Times, 1969), 12–14.

30. See Mansfield, *That Fateful Class,* 256–63.

31. Morris, *Reading, 'Riting, and Reconstruction,* 92; U.S. Commissioner of Education, *Annual Report, 1891–92* (Washington, DC: GPO, 1892), 864–65, 1234–37; Glenda E. Gilmore, *Gender and Jim Crow: Women and the Politics of White Supremacy in North Carolina, 1896–1920* (Chapel Hill: University of North Carolina Press, 1996), 31–47. See also, Christine Ogren, *The American State Normal School: An Instrument of Great Good* (New York: Palgrave Macmillian, 2005); James Fraser, *Preparing America"s Teachers: A History* (New York: Teachers College Press, 2007).

32. Robert Sherer, *Subordination or Liberation? The Development and Conflicting Theories of Black Education in Nineteenth Century Alabama* (University, AL: University of Alabama Press, 1977), 6–7; Edmund O'Neal, "The Rise and Development of State Teacher Training for Negro Elementary Teachers in Alabama" (master's thesis, University of Cincinnati, 1939), 28–30.

33. Horace Mann Bond, *Negro Education in Alabama: A Study of Cotton and Steel* (1939; repr., New York: Atheneum, 1969), 105–10; "Special Preliminary Announcement for 1949–50": page 3, H. Councill Trenholm Collection, box 21, College Program/Reports, folder 9, Alabama State University Archives and Special Collections (hereafter ASU Archives), Montgomery, Alabama.

34. See Louis R. Harlan, *Booker T. Washington: The Making of a Black Leader, 1856–1901* (New York: Oxford Univ. Press, 1972), 166–68; Sherer, *Subordination or Liberation?* 10–13, 27–28; "The Diamond Jubilee Anniversary," page 1-2, ASU Pamphlets, box 1, ASU Archives; H. Councill Trenholm, "Some Background and Status of Higher Education for Negroes in Alabama," 1949 *ASTA Association Yearbook,* Harper C. Trenholm Collection, box 23, folder 3, ASU Archives.

35. W. T. B. Williams, "State Normal School, Montgomery, Alabama," May 10, 1906, General Education Board Collection, Series 1, Subseries 1, box 13, folder 101, Rockefeller Archive Center; *An Educational Study of Alabama,* U.S. Bureau of Education Bulletin, no. 41 (Washington, DC: GPO, 1919), 196, 380; "Catalogue of the Faculty and Students for 1922–23 and Announcements of Courses for 1923–24," Harper C. Trenholm Collection, box 21 College Programs/Reports, folder 9, ASU Archives; Sherer, *Subordination or Liberation?* 25–26.

36. "In Memorial Tribute to its Fourth President," George Washington Trenholm Collection, box 1, folder 1, ASU Archives; Naomi Webb, "The Life of George W. Trenholm," *The State Normal Journal* 1 (March 1928): 22–23, ASU Pamphlets, box 2, ASU Archives; "Notes: Improving the Rural Teaching Force Through Teachers Institutes," Harper C. Trenholm Collection, box 5, ASU Archives.

37. *Catalog, 1920–21,* 17–18, 31, ASU Archives; "Announcement of Curricula with Descriptions of Courses and Certification Regulations for 1930–31," 2; and *The Faculty of 1934–35,* ASU Archives and Special Collections, box ASU Files—Catalogues; "Five Years of Junior College Work at State Normal," *Fifty-fourth Commencement of The State Normal School, Montgomery, Alabama, Wednesday, May 27th, 1925,* ASU Archives, ASU History, 1921–1925—box 6.

38. "Five Years of Junior College Work at State Normal"; "Catalogue of the Faculty and Students for 1922-1923 and Announcements of Courses for 1923-24," 23, Harper Council Trenholm Collection, box 21 College Programs/Reports, folder 9, ASU Archives.

39. "The 1948 Summer Quarter Announcement" and "The 1950 Summer Quarter Announcement," Harper C. Trenholm Collection, box 21 College Programs/ Reports, folder 9, ASU Archives; "The Faculty of 1935—1936," 34 and "Annual Catalogue, 1937—38," 78, ASU Archives, ASU Files—Catalogues.

40. In the 1929–30 school year, the college's overall enrollment of 5,004 was distributed as follows: summer school, 2,369; extension, 1,437; high school, 640; college, 404; grades 1–9, 154. See "The Faculty of 1935—36," 34, 36; "The Graduates Book of the 1937 Summer Quarter," 5, 11; "Presenting . . . The Graduates book for the 1938 Summer Quarter," 3; "The Graduates Book for the Summer Quarter 1939," 3, all in ASU Archives and Special Collections, ASU Files—Catalogues.

41. Alabama State in 1937–38," no page numbering, ASU Archives, box, ASU Files—Catalogues; "The 1950 Spring Quarter Announcement," cover page, and "The 1952 Spring Quarter Commencement," 12, Harper C. Trenholm Collection, box 21 College Programs/Reports, folder 9, ASU Archives; "Explanatory and Financial Report on In-Service Training for Negro Teachers in Rural Schools at Snow Hill Institute, Snow Hill, Alabama," cover page and pages 1–6, General Education Board Collection, Series 1, Sub-Series 1, box 18, folder 153, Rockefeller Archive Center.

42. Letter from G. W. Trenholm to John W. Abercrombie, April 28, 1922, George Washington Trenholm Collection, box 1, folder 4, ASU Archives; "Catalog, 1920—21," 13–14; *The State Normal Courier-Journal* 4 (January 23, 1926): 8, ASU Pamphlets, box 2, ASU Archives; "Catalogue of the Faculty for 1929–30 and General Announcement for 1930–31," 23, ASU Files—Catalogs; "Announcement of Curricula with Description of Courses and Certification Regulations for 1930—31," 64–65; "The Faculty of 1934—35,"; "The Faculty of 1935—36," 32–33; "Pre-Catalogue Announcements for 1947–48," 21, Harper C. Trenholm Collection, box 21, College Programs/Reports, folder 9, ASU Archives.

43. Trenholm, "Some Background and Status of Higher Education for Negroes in Alabama," 41—49.

44. "Pre-Catalogue Announcements for 1947—48," 10, 21, 24, 57–58; "The 1952 Spring Quarter Commencement," 12, ASU Archives

45. See Felton G. Clark, *The Control of State Supported Teacher-Training Programs for Negroes*, Contributions to Education #605 (Bureau of Publications, Teachers College, Columbia University, 1934). Clark misses six institutions, three teachers colleges and three municipal colleges. See also, Patrick J. Gilpin and Marybeth Gasman, *Charles S. Johnson: Leadership Beyond the Veil in the Age of Jim Crow* (Albany: State University of New York Press, 2003); Marybeth Gasman, "Rhetoric vs. Reality: The Fundraising Messages of the United Negro College Fund in the Immediate Aftermath of the Brown Decision," *History of Education Quarterly* 44 (Spring 2004): 70–94; Joy Ann Williamson, "'This Has Been Quite a Year for Heads Falling': Institutional Autonomy in the Civil Rights Era," *History of Education Quarterly* 44 (Winter 2004): 554–76; Callie L. Waite, *Permission to Remain Among Us: Education for Blacks in Oberlin, Ohio, 1880–1914* (Westport: Praeger, 2002); Margaret Smith Crocco and Cally L. Waite, "Education and Marginality: Race and Gender in Higher Education, 1940–1955," *History of Education Quarterly* 47 (Spring 2007): 69–91.

46. See "The Meaning of Progress," in *The Souls of Black Folk*, W. E. B. DuBois (New York: New American Library, 1969), 96–108; see also Mansfield, *That Fateful Class*, 191–93; Higginbotham, *Righteous Discontent*, 37, 96–108; Gilmore, *Gender and Jim Crow*, 35.

47. Hoy Taylor, "An Interpretation of the Early Administration of the Peabody Education Fund" (PhD diss., George Peabody College for Teachers, 1933); E. B. Robert, "The Administration of the Peabody Education Fund from 1880 to 1906" (master's thesis, George Peabody College for Teachers, 1936); "Report of the Board of Trustees of the State Normal Schools to the Legislature of Alabama, January 1915," Harper C. Trenholm Collection, box 21 College Programs/Reports, folder 12, ASU Archives; Leo Favrot, "Report on Summer Schools for Negroes, 1931," General Education Board Papers, box 302, folder 316, Rockefeller Archive Center; "Information Concerning Board Aid for Summer Schools for Negroes (1940)," General Education Board Papers, box 301, folder 315, Rockefeller Archive Center.

48. Carter G. Woodson, *The Rural Negro* (New York: Russell and Russell, 1969), Table 11, page 214.

49. See Chirhart, *Torches of Light*, 75–91, for a discussion of these issues in Georgia.

50. "Examinations for Colored Teachers in Virginia," *Southern Workman* 38 (April 1909): 196; John M. Gandy, "Fifty Years of Professional Growth of Negro Teachers in Virginia," *Virginia Teachers Bulletin* 15(January 1938): 9–10; "To the State Board of Education, Richmond, Virginia," W. T. B. Williams Collection, box 10 Misc. Reports, Tuskegee University Archives, Tuskegee, AL; *An Educational Study of Alabama*, U.S. Bureau of Education, Bulletin 1919, no. 41 (Washington, DC: GPO, 1919), 380, 399.

51. Catalogue 1914–1915, quoted in Fancher, "Tennessee State University," 30.

52. Tennessee Agricultural and Industrial Normal School *Bulletin*, 3 August 1914, 62–63; George W. Brooks, *History of the Tennessee Education Congress, 1923–1967* (Washington, DC: National Education Association, 1975), 11–14. For the most comprehensive study to date on the Rosenwald schools and on black teachers' contributions, see Mary S. Hoffschwelle, *The Rosenwald Schools of the American South* (Gainesville: University Press of Florida, 2006).

53. Michael Fultz, "Caught Between a Rock and a Hard Place: The 'Displacement' of Black State Teachers Associations, 1954–1971," unpublished paper presented at the Annual Meeting of the American Educational Research Association, Chicago, April 2007; Vernon McDaniel, *History of the Teachers State Association of Texas* (Washington, DC: National Education Association, 1977); Gilbert Porter and Leedell Neyland, *The History of the Florida State Teachers Association* (Washington, DC: National Education Association, 1977); Thelma D. Perry, *History of the American Teachers Association* (Washington, DC: National Education Association, 1975); Darlene C. Hine, "Black Professionals and Race Consciousness: Origins of the Civil Rights Movement, 1890–1950," *Journal of American History* 89 (April 2003): 1279.

54. J. Rupert Picott, *History of the Virginia Teachers Association* (Washington, DC: National Education Association, 1975), 221–22.

55. Michael Fultz, "The Displacement of Black Educators Post-Brown: An Overview and Analysis," *History of Education Quarterly* 44 (Spring 2004): 11–45; Adam Fairclough, "The Costs of Brown: Black Teachers and School Integration," *The Journal of American History* 91 (February 2004): 43–55; Adam Fairclough, *A Class of Their Own: Black Teachers in the Segregated South* (Cambridge: Harvard University Press, 2007); Carol F. Karpinski, "Bearing the Burden of Desegration: Black Principals and Brown," *Urban Education* 41 (November 2006): 237–76.

56. Vanessa Siddle Walker, *Their Highest Potential: An African American School Community in the Segregated South* (Chapel Hill: University of North Carolina Press, 1996); Vanessa Siddle Walker, "Valued Segregated Schools for African American Children in the South, 1935–1969: A Review of Common Themes and Characteristics," *Review of Educational Research* 70 (Fall 2000): 253–85. See also, David Cecelski, *Along Freedom Road: Hyde County, North Carolina and the Fate of Black Schools in the South*

(Chapel Hill: University of North Carolina Press, 1994); Adah Ward Randolph, "The Memories of an All-Black Northern Urban School: Good Memories of Leadership, Teachers, and the Curriculum," *Urban Education* 39 (November 2004): 596–620; Michele Foster, *Black Teachers on Teaching* (New York: New Press, 1997).

57. Barbara Shircliffe, "'We Got the Best of that World': A Case for the Study of Nostalgia in the Oral History of School Segregation," *Oral History Review* 28, no. 2 (Summer/Fall 2001): 60; Jack Dougherty, "From Anecdote to Analysis: Oral Interviews and New Scholarship in Educational History," *Journal of American History* 86 (September 1999): 712–23.

58. There is no comprehensive history of the Jeanes supervising teachers (1908–1968), although both then-contemporaries and present-day historians generally consider them as exemplary models of engaged African American educators during the era of de jure segregation, epitomizing dedication to the improvement of black schooling through a combination of in-service and community-building activities. For informative overviews, see Valinda Littlefield, "'To Do the Next Needed Thing': Jeanes Teachers in the Southern United States, 1908–1934," in *Telling Women"s Lives: Narrative Inquiries in the History of Women"s Education*, Kathleen Weiler and Sue Middleton (Buckingham: Open University Press, 1999), 130–45; Littlefield, *I Am Only One, But I Am One*; see also Gilmore, *Gender and Jim Crow*, 160–65, for a discussion of how Jeanes teachers worked to turn the initial industrial education component of their work "into a self-help endeavor."

59. See, for example, Charles S. Johnson, *Growing Up in the Black Belt: Negro Youth in the Rural South* (Washington, DC, American Council on Education, 1941); Charles S. Johnson, *Shadow of the Plantation* (Chicago, University of Chicago Press, 1934); Charles S. Johnson, "The Negro Public Schools: A Social and Education Survey," Section 8 of *Louisiana Educational Survey* (Louisiana Educational Survey Commission, 1942).

60. Charles Thompson, "Negro Teachers and the Elimination of Segregated Schools," *Journal of Negro Education* 20 (Spring 1951): 139.

61. See Adam Fairclough, *Teaching Equality*, 42–47; Fairclough, *A Class of their Own*, 355–91; Littlefield, *I Am Only One*, 7–8; Chirhart, *Torches of Light*, 6.

62. Adam Fairclough, *Teaching Equality: Black Schools in the Age of Jim Crow* (Athens, GA: University of Georgia Press, 2001). See also, Vanessa Siddle Walker, "Organized Resistance and Black Educators' Quest for School Equality, 1878–1938," *Teachers College Record* 107 (March 2005): 335–88.

63. Higginbotham, *Righteous Discontent*, 42. V. P. Franklin has promoted a potentially very significant line of research on "Cultural Capital and African American Education." See special issue of *The Journal of African American History* 87 (Spring 2002).

CHAPTER 5

AMERICAN PUBLIC SCHOOLING AND EUROPEAN IMMIGRANTS IN THE EARLY TWENTIETH CENTURY: A POST-REVISIONIST SYNTHESIS

Michael R. Olneck

Historians are important mythmakers.[1] Among the central legends of American history is that of the immigrant and the school. The myth that—through schooling—early twentieth-century European immigrants to the United States were afforded and embraced unparalleled opportunities to achieve social mobility and to "become American," has shaped responses to persisting poverty among African Americans, informed contemporary education policy toward "English Language Learners," and, generally, stood as an object lesson for how success in America is available to all.[2] Historians, as John Bodnar has observed, have contributed to that myth by depicting immigrants as "cherishing the idea of free public education and the promise it offered for social success," and as demonstrating a "'commitment' to the American dream of personal advancement through schooling."[3]

For radical revisionist historians, the "immigrant story" of schooling, opportunity, and meritocracy was among the important myths of American education to be debunked. Colin Greer, especially, devoted his efforts to demolishing the "Great School Legend" that depicts American schools as having "accommodated, assimilated, and set on the road to . . . success" the mass of foreigners entering the United States before World War I.[4] That story, according to Greer, is "staggeringly exaggerated."[5] Rather, Greer argued, the urban immigrant poor experienced widespread failure in American schools, left school as early as possible, and experienced whatever modest social mobility they enjoyed in spite of—not because of—schooling.

To Greer, scholars like Lawrence Cremin—who regarded the efforts undertaken by schools to "Americanize" immigrant children as part of a broader progressive education reform animated by humanitarian concerns—confused the rhetoric of reformers with reality, and failed to see that far from offering opportunity to immigrants, schooling was an apparatus designed to subject immigrants to control, and to safeguard society's existing social and economic hierarchies.[6] "The school reform movement," Greer wrote, "stood solidly in defense of the white Anglo-Saxon Protestant dominance."[7] In a similar vein, Joel Spring interpreted the expansion of functions undertaken by early twentieth-century schools to Americanize immigrant children as part of the institutional elaboration of a new "corporate order," in which schools implemented organizationally novel forms of social control.[8] Writing of an earlier period, Michael Katz took note that widespread anxieties about the threat to cultural, social, and economic order seemingly posed by large-scale immigration "propelled the establishment of systems of public education; from the very beginning public schools became agents of standardization." [9] Katz attributed the later defeat of "democratic localism," and the triumph of bureaucratically organized school systems to "a gut fear of the cultural divisiveness inherent in the increasing religious and ethnic diversity of American life."[10] In their landmark publication, *Schooling in Capitalist America*, Samuel Bowles and Herbert Gintis emphasized the school's role in integrating early twentieth-century urban working-class immigrant children into a highly stratified workplace by channeling them into non-academic paths leading, ultimately, to places at the bottom of the capitalist system of production.[11]

Other revisionist historians, adopting frameworks less based on assumptions of elite domination or class conflict than those adopted by their colleagues, emphasized the perceptions of social dislocation, anxieties, and uncertainties animating educators' responses to late nineteenth- and early twentieth-century immigration. Marvin Lazerson emphasized turn-of-the-century educational leaders' turn to the schools to constrain the disruption and fragmentation of an increasingly urban, industrial, heterogeneous society. In these educators' views, society could no longer rely on the simple inculcation of common principles to cultivate shared outlooks and recognizably American ways of living.[12] In Lazerson's view, the expansion of the school into kindergartens, manual education (and later, vocational education), and civic education was to be explained by the beliefs of educators that explicit tutelage in social, political, and economic behavior was now necessary to repair a social fabric experiencing unprecedented strain. For these educators—who idealized an idyllic, harmonious, and natural rural past—the need to counteract the ills of the urban environment and the inadequacies of immigrant families were palpable reasons for the transformation of the school from an institution continuous with family and community, to one that sought deliberately and systematically to reconstruct family and community life.

David Tyack, like Lazerson, emphasized the role of late nineteenth- and early twentieth- century educators' concerns about intensified social conflict and threats to social stability in shaping education reform.[13] Tyack, too, highlighted the priority educators assigned to weaning immigrant children from morally corrupting homes and communities, and to systematically training them in industry, temperance, and obedience. More so than others, Tyack situated educators' responses to the large and persisting influx of immigrant children into American cities, within the context of the practical problems those children posed to school administrators and teachers working under enormous new pressures. For that reason, Tyack—though acknowledging the ethnic, racial, and social class stratification that emerged in public schools—was less condemnatory of the testing and classifying practices schools adopted than were some other revisionists. Similarly, Tyack—while recognizing the ethnocentrism with which educators approached immigrant students and parents, and the shame at being "foreign," they sometimes inflicted on immigrants—nevertheless acknowledged the altruism with which they approached their work. The "rhetoric of Americanization," he wrote, was often messianic—a mixture of fear outweighed by hope of a desire for social control, accompanied by a quest for equality of opportunity for the newcomers under terms dictated by the successful Yankee."[14]

Tyack also recognized that whatever the aspirations of the schools for immigrants, the reach of the school was limited, and immigrant children, families, and communities were not powerless in how they appropriated what the schools offered, and in politically pressing for pedagogical and curricular accommodations like the incorporation of non-English languages as languages of instruction or subject matter. Finally, while attentive to important variations among groups in their embrace of public schooling, Tyack rejected arguments that schooling was imposed upon resistant immigrant communities, and concluded that "no brigades of attendance officers could have coerced such masses of children into school if their parents had strongly opposed public education."[15]

Subsequent historical scholarship concerning American schooling and immigrants has, by and large, validated the interpretations advanced by Lazerson and Tyack thirty years ago. At the same time, it has—in some cases—achieved further conceptual and empirical richness, and attained greater analytical complexity. Here I will attempt to synthesize the work of early critics of revisionists' accounts of immigrants and schooling, and of more recent scholarship, in order to better understand and explain the purposes, policies, and practices with which educators approached early twentieth-century European immigrants. I will also examine ways in which immigrants pursued their own political, social, and economic purposes through schooling, all areas of inquiry that early revisionists neglected. Finally, I will call attention to the ways in which social historians have recently recast debates about the role of "culture" and "structure" in the analysis of immigrant school attainment and social mobility through schooling.

EDUCATORS' PURPOSES, POLICIES, AND PRACTICES WITH RESPECT TO IMMIGRANTS

Critics of Katz's *Class, Bureaucracy, and Schools* and Karier, Violas, and Spring's *Roots of Crisis* objected to some revisionists' inclination to infer motives and purposes from the eventual outcomes of education reforms.[16] They were more prepared to credit reformers and educators with having the best interests of immigrant and poor students and communities at heart.[17] Irrespective of reformers' and educators' motivations, making sense of *how* they shaped schools to respond to the influx of immigrants requires understanding their ideological presuppositions and value commitments, as well as recognizing broader trends into which the schooling of immigrants fit.

Early twentieth-century education reformers were just that, *reformers*. They sought to mitigate the socially destructive effects of the market revolution, and to (re)construct an idealized "republican liberal" society of striving, personally responsible, like-minded, and public-regarding individuals. They sought a society in which the degrading impacts and social fragmentation of the urban slum would be diminished. They aspired to rationalize the market society, not to either reproduce or repudiate it. Their approaches, in part harking back to common school presuppositions about the potential of schooling to further moral education—as well as to established democratic ideals about individual opportunity in America and long-held assumptions that schooling should prove useful—sought to expand and intensify the scope of the school, to legitimize its broader authority, and to elaborate its internal workings. Immigrants—because they were massively numerous and territorially concentrated, severely poor, often exploited, and foreign—constituted a threat to the moral and social order reformers sought to protect, and, therefore, posed a distinctive challenge to their efforts.[18]

In the view of early twentieth-century education reformers, the homes, families, and communities from which immigrant children came were deficient in providing moral guidance, social constraints, and the skills needed to participate in modern, complex, and democratic American life. With faith in the capacity of institutions to successfully order social life, to accomplish moral uplift, and to impart socially valuable skills—and drawing on, and going beyond, the precedents established by the activities of the settlement houses (*e.g.*, kindergartens, playgrounds, mothers' clubs, domestic and trades classes, English-language instruction, home visitations)—education reformers turned to the schools to serve as "a central agent in the transformation and assimilation of immigrant aliens into the logic and lifeways of the dominant culture."[19] While undoubtedly shaping schools to reproduce the value of dominant cultural capital, and to subordinating immigrants to the verdicts of the dominant culture, education reformers approached the schooling of immigrant children with a view to "sharing" the opportunities and benefits of

American life. Schools were to become part of wider processes "by which the foreign population would be modified, elevated, and reformed."[20]

The deliberate use of schooling to "Americanize" immigrant students was one aspect in a more general expansion and intensification of schooling as a means of civic socialization for all youth. Under capitalist industrialization, the workplace had ceased to be viewed as an effective locus of socialization and source of social cohesion, and reformers invested in schooling with new significance as a site for cultivating democratic citizenship.[21] Civic socialization, however, was inextricably linked with, and sometimes indistinguishable from, socialization for the workplace, and the values and dispositions that schools in the early twentieth century promoted were not altogether different from those promoted in nineteenth-century schools. Superintendent William Wirt's efforts in Gary, Indiana, for example—to ensure that schools promoted cooperation, industriousness, thrift, temperance, cleanliness, patriotism, punctuality, self-discipline, self-reliance, and respect for authority among immigrant and native students alike[22]—may have differed in methods from the efforts of his nineteenth-century predecessors, relying less on didactic methods and more on the social organization of students' experience, but they did not differ a great deal in purpose.

While historians have readily recognized that American schools have been grounded in—and have attempted to promote—native, middle-class values and world-views, they have only recently recognized the significance of long-term cultural transformations in middle-class views of childhood and personhood for the schooling of immigrants. The lengthening of the period of formal schooling, and the organizational elaboration and standardization of the schooling process characteristic of the late nineteenth and early twentieth centuries, were impelled, in part, by a broad cultural trend within the middle-class toward sentimentalizing children and childhood, and re-defining "growing up" to include a lengthy period of formal preparation and semi-autonomy.[23] Schooling, Stephen Lassonde wrote, introduced a "powerful new element in the socialization of children, for schooling defined and regulated childhood and youth as discrete, sequenced phases of preparation for adulthood."[24] A significant aspect of this transformation was an increased emphasis on the futures and preferences of individual youths, a process Paula Fass has described as "individualizing destiny."[25] This emphasis clashed with the views of many immigrant parents on the utility of offspring for the family economy and on the value of work, and with deep values concerning the obligations of offspring to their parents. These included the obligation to recognize the primacy of parental authority, as well as to be readily available in times of family hardship. Schooling, consequently, threatened to induce conflict and ambivalence into immigrant families, undermine the "development of family morality," and "breaching the continuity of culture that the immigrant habits tried to preserve."[26]

"Individualizing destiny" carried with it the implication of differentiating the educational experience of students, while, at the same time, it aggravated the problem of promoting common allegiances and feelings of social solidarity among diverse students. The invention of extracurricular activities, assemblies, whole-school social activities like dances, and the elevation of inter-scholastic sports were intended as antidotes to the centrifugal effects of growing academic differentiation. Extracurriculars became, in Paula Fass's words, "the repository for the old common school ideal."[27] Academic differentiation in the form of formal tracking, based on newly-developed standardized tests that enacted widening expert and popular belief in IQ, was, in the view of education reformers, a scientifically warranted and democratic response to the "individual needs" of increasingly heterogeneous student bodies. As implemented in the schools, however, tracking practices served goals of administrative efficiency more than they did the welfare of many students. Immigrant students, in particular, were often identified as less able, and precluded from academically valuable opportunities.

The practices of the schools in response to the influx of immigrant students in the early twentieth century, cannot, however, be understood solely as the implementation of political or educational theory, though both formalized theory and shared ideologies made certain responses more sensible and natural than others. During the 1890s and for several decades afterward, teachers in urban schools increasingly confronted large classes of students whose fluency in English, past academic preparation and performance, interest in schooling, and intentions for the future varied enormously, making instruction far more difficult than in the past. Well before IQ testing made its appearance, schools were engaged in grouping students according to criteria intended to simplify the tasks of teaching. In New York City, for example, in 1898, District Superintendent Julia Richman began experimenting with "bright," "medium," and "poor" groups at the same grade level. Later, Richman instituted experimental "C" classes, or "steamer," or "vestibule" classes for learning English; "D" classes for kids going to get their working papers at a permissible leaving age; and "E" classes for rapid advancement groups, intended for late-entrants who were academically adept and could move quickly, thus leaving classrooms for which they were over-aged.[28] While not initially widespread, these experiments demonstrate the significance for the transformation of the schools during the progressive period, of on-site responses to practical problems, and the role of the immigrant presence in prompting those transformations.

Similarly, the adoption of vocationally- and practically-oriented curricula in urban high schools was viewed by educators as necessary to engage, retain, and meet the needs of increasingly diverse student bodies. For example, around 1913, New York's Washington Irving High School had a housekeeping flat, and taught courses on marriage, baby care, personal hygiene, household sanitation, and first aid. "All in all, this kind of high school was a direct attempt to meet the most obvious needs of the immigrant adolescent girls; to equip them for jobs, marriage and urban living, while raising their standards and aspirations."[29] This benign interpretation of increasing differentiation of schooling experiences does not gainsay the stratifying effects by class, race, and ethnicity that tracking and vocationalization produced, but it does caution us against confusing consequences with purposes.

While varying in their appreciation or disdain of immigrant children and their cultural practices, teachers in urban schools almost universally viewed it as their obligation to prepare immigrant children to live according to American cultural norms. Julia Richman, for example, prohibited the use of Yiddish in the schools she superintended, and instructed teachers to assign demerits for its use, even when on the playground or in the bathrooms.[30] Teachers constantly informally admonished students on the importance of cleanliness and politeness, and attempted to imbue them with native, middle-class tastes, values, ethics, and conceptions of civic duty, as well as promoted these ends in newly adopted courses in such subjects as "Course of Study in Manners and Conduct of Life."[31] They did this under the rubric of formal curricula based on principles such as those embodied in a 1903 New York City school curriculum that included inculcating the love of "good literature," singing "high class" music, and learning history as an introduction to the American "heritage."[32]

IMMIGRANTS' RESPONSES TO THE SCHOOLS

One question left unaddressed by early revisionist analyses of the role of the schools in the acculturation of immigrant youth is to what extent these efforts were welcomed, ignored, deflected, resisted, or, perhaps most importantly, appropriated to purposes and functions unanticipated by educators. Critics challenging the revisionist model of "cultural imposition" have rejected the depiction of "nonelites as powerless victims of unwanted educational arrangements and as acted-upon subjects of imperial tutelage,"[33] and have called for research that, in contemporary parlance, examines the "agency" of immigrant youth, families, and communities. Ronald Cohen and Raymond Mohl, for example, concluded from their study of Gary, Indiana schools that, in contrast to revisionist and traditionalist interpretations, "immigrants had a large degree of control and self-determination when it came to educational institutions. "[34] Stephen Brumberg reminded his readers not to confuse the message *transmitted* with those internalized by students.[35] Paula Fass argued that "outsiders," including immigrants, have "often through schooling, redrawn the boundaries of the culture which had initially defined them apart."[36] Various outsider groups, Fass argued, have not necessarily acted as educators and the larger public have expected them to. Rather, they have used schooling to achieve their own goals, and have done so in ways that did not conform to officially approved expressions of pluralism.

The response of immigrant communities themselves to the "Americanizing" efforts of the schools was mixed, and by no means necessarily resistant.[37] In New York, Stephen Brumberg has concluded, the schools' Americanizing curriculum was a success "due in large part to the active collaboration of key elements in the City's Jewish community."[38] Jewish immigrant parents, no less than educators, favored basic scholastic preparation that emphasized English literacy, as well as acculturation to accommodate children to American norms. Selma Berrol attributed noticeable immigrant Jewish success in New York schools to Jewish parents' acceptance of the

Americanization efforts of the school.[39] Jews relied on the extensive provision of supplementary religious schooling to ensure that Jewish identity and commitment were sustained, even as their youth availed themselves of the opportunities provided by public schooling.[40] Moreover, measures that offended immigrants' sensibilities could be evaded. For example, in Gary, Indiana, immigrant children ignored the release-time program for religious education initiated by Superintendent William Wirt, and, around 1910, when Wirt refused to institute night school classes using the Polish language, attendance fell off significantly.[41] More generally, immigrants could simply ignore the Americanization classes so earnestly provided by educators.[42]

The desire of some immigrant adults for night school classes provided in their own language should not be confused with a widespread desire on the part of immigrant communities to utilize the public schools for linguistic preservation. While immigrant and ethnic intellectuals, writers, editors, and publishers often pressed schools to incorporate homeland languages into their curricula,[43] and while community groups sometimes submitted petitions in support of such demands or requested that public school classrooms be made available for after-school homeland language classes[44]—with the exception of German-language classes in a limited number of midwestern cities in the late nineteenth and earlytwentieth centuries[45]—the responses of the mass of ordinary immigrant parents and students to these opportunities were short-lived and meager, with classes often languishing for lack of enrollment.[46] Rather than homeland languages in the public schools being indicative of immigrants' rejection of Americanization, the pursuit of the inclusion of such languages was part of the processes by which "ethnic groups" were formed, and incorporated into American local politics. The offering of homeland languages as subjects of study represented recognition on the part of public authorities that a group had become a respected part of the American body politic.[47]

At times, immigrants' opposition to the projects of education reformers took the form of organized political mobilization and protest. New York's eastern European Jews, for example, who wanted to ensure that their children remained academic,[48] collectively, and, in cooperation with other political groups, successfully resisted the adoption of the Gary Plan during the mid-1910s.[49] Similarly, between 1917 and 1930, ethnic communities in Buffalo, New

York, resisted the establishment of junior high schools, which parents felt would discourage their children from continuing on to high school.[50]

Not all immigrant families or communities found the public schools congenial to their aspirations. In the late nineteenth century and pre-1930s twentieth century, for immigrant communities that were resistant to anti-Catholic nativist undercurrents and the Americanizing and secular agendas of public schools, and were committed to cultural- and religious-based schooling, attendance at parochial schools was an often available alternative. In particular, in national parishes associated with specific ethnic communities, attendance at such schools was, for many, a natural and unreflective path.[51] Indeed, John Ralph and Richard Rubinson have concluded that between 1890 and 1924, mass immigration accelerated the rate of growth only of parochial schools, and not public schools.[52] Though, competition from parochial schools impelled some districts to disproportionately locate public schools within attendance areas served by the Catholic schools.[53] In some cities, attendance at Catholic schools represented over 40 percent of Catholic students, and was particularly strong among Poles and other Slavic communities who initially looked to schools more for moral and cultural purposes than for goals of economic advancement.[54]

Parochial schooling proved, however—in historical perspective—to be integrative, not separatist institutions. "National" parishes brought together families of diverse regional and local origins, whose dialects often differed, and provided the institutional community through which they might become "German" or "Polish" or some other "nationality," in short, to become American ethnic groups.[55] Over time, the curriculum and pedagogy of ethnically distinct parish schools became increasingly Americanized, and Catholic schools became an English-monolingual path of incorporation into American society based on religious identification, not on national origin. [56]

Schooling may well have proven more potent in teaching immigrant youth what "being American" meant than it did in readily transforming them into "Americans," or, as revisionist scholars emphasize, workers. Quite apart from the emergence of tracking and vocational curricula in the high schools, immigrant youth required only the models of their own parents' lives, and their own episodic experiences of job-holding to envision their futures as "workers."[57]

Schools provided immigrant youth with models of middle-class American lifeways and new sources of authority, while reminding them of the social distance and boundaries between the worlds they and the "Americans" inhabited. In institutionalizing a trajectory of individual development, schools demonstrated an alternative to the matrix of obligations characteristic of the morality of the immigrant family economy, but only slowly did youth from most immigrant communities come to pursue that alternative, through remaining in school beyond the age of compulsory attendance. Once they did, however, high schools in particular became sites in which to participate in changing cultural practices characteristic of American youth culture, including consumerism, pursuit of popularity, and unsupervised dating.[58]

In treating "immigrant" and "American" as a dichotomy, there is a danger of overlooking a central contribution of schools to immigrant incorporation, namely, the role of schools as important sites in the elaboration of *ethnic* identities and associations. While immigrant youth might learn much of how to be Italian, Polish, Greek, or Jewish, in America—through life within their own families and communities—they learned and developed the meaning of ethnic identities and associations in relation to others with whom they had contact in the schools.[59] In ethnically segregated elementary schools, the contrasts with teachers and staff were no doubt most salient. In more heterogeneous high schools, interactions with, and contrasts between, themselves and native American peers took on greater salience.[60] Of significance to broader processes of class formation, in which the urban working class was fragmented and (dis)organized along lines of ethnicity,[61] socioeconomic differences between themselves and native American students were experienced by immigrant youth as "ethnic."[62]

Extracurricular activities in particular proved an important locus for the shaping of ethnic identities and relationships, at the same time that they provided opportunities to follow American norms of crafting individual identities. Contrary to the hopes of educators that extracurricular activities would mix students according to interests presumed to be independent of ethnic group membership, Paula Fass found, as I have noted elsewhere, "pronounced ethnic differences, particularly among young men, in patterns of extracurricular participation in New York City's high schools during the 1930s and 1940s. More importantly, the nature of those differences varied from school to school, depending upon patterns of social class and ethnic composition. This latter variation suggests strongly that the dynamics of ethnic participation were not simple extrapolations of traditional affinities, but arose out of context-specific interaction between groups [competitively] seeking a place within each school's prestige and status hierarchies."[63] Extracurricular activities were by no means ethnically homogeneous, but they were differentiated in ways that intergroup contact was mediated by ethnic bonds and by school-to-school particularities of stratification and composition. "Ironically," Fass concluded, contrary to the idealizations of Americanizers, "the high school not only did not destroy the preschool associations of students, it encouraged, supported and thereby strengthened them."[64] Thus, the schools were one site in the broader processes of immigrant incorporation that are best comprehended not as "assimilation," but as acculturation, accompanied by the retention and elaboration of ethnic identities.[65]

IMMIGRANT POLITICAL INFLUENCES IN EDUCATION

As noted in the case of successful resistance by immigrant groups in some cities to imposition of the Gary Plan on junior high schools, immigrants were not powerless in defending their interests in educational politics. In the case of the incorporation of ethnic languages into the public schools as either languages of instruction or subjects of study, they demonstrated success, as well, in positively pursuing their interests.

The case of ethnic language incorporation into public schools with which historians are most familiar is, of course, the incorporation of German from the post-Civil War period until World War I.[66] Less well-known are the successful efforts to introduce Hebrew, Italian, and Polish into the public schools of some major American cities.[67] Interestingly, only negligible numbers of ethnic students enrolled in these courses.[68] What is most significant about these efforts is that they were important parts of the processes by which American ethnic groups were formed, consolidated, mobilized, and incorporated into the matrices of American local politics.[69] Moreover, insofar as schools perform a "status-conferring function",[70] inclusion of these languages in the public schools represented recognition on the part of public authorities that a group had become a respected part of the American body politic.[71]

The success of immigrants and their descendants in using the schools to secure recognition of ethnic group status needs to be seen, however, in the context of the separation of home and community from the workplace, and the consequent dispersion of working class power in educational politics.[72] In the view of Katznelson and Weir, "the representation of society in ethnic terms imposed limits on what could be said and fought about in political life."[73] The result was not, however, a structure of schooling determined by capitalist power, as Bowles and Gintis claim. Rather, the result is more that "the structure of schooling in the United States reflects the absence of class effects rather than their presence."[74]

IMMIGRANT SCHOOL SUCCESS AND THE USE OF SCHOOLING FOR SOCIAL MOBILITY

The degree to which European immigrants experienced academic success in American schools has received considerable attention from historians and social scientists. At issue in this work are questions of the truth of mythology celebrating immigrant social mobility via the schools, claims about ethnic variation, theoretical conceptualizations of "culture" and "social structure," and the nature of immigrants' strategies with respect to education, work, and family.

In their attempts to address these questions, scholars have synthesized published data from the first third of the twentieth century,[75] and have quantitatively modeled individual-level historical census data.[76] Their work aims to determine whether there are effects of immigrant ethnicity on education outcomes that cannot be accounted for by prior or contemporaneous measures of socioeconomic standing, literacy, family size, and the like. If there are, we can conclude that schooling provided at least some groups with routes of mobility that exceeded group members' opportunities provided by their families' socioeconomic position. Such net "ethnic effects" have been problematically attributed to the influence of "culture," while the effects of socioeconomic variables have been attributed to "structure." Of greatest credibility are the studies utilizing individual-level data, the most valuable of these being Joel Perlmann's longitudinal study of data for individuals from Providence, Rhode Island, covering the period 1880–1930.[77]

Perlmann analyzed data linked from local and state school and census records, and the manuscript U.S. Census. He found, as expected, that the children of middle-class families were much more likely to attend high school than were the children of working-class families, and that differences in socioeconomic standing contributed substantially to accounting for apparent differences in school persistence among ethnic groups. But, at the same time, he also found that important ethnic differentials persisted even when socioeconomic differentials were controlled. In 1880, the offspring of Irish immigrants and of Irish-background native-born parents were substantially less likely to enter high school than the offspring of "Yankees," even when their socioeconomic origins were comparable. But, by 1900, this disadvantage had completely disappeared among socioeconomically comparable individuals. At the same time, Perlmann found large and persistent disadvantages in high school entrance among Italian youth, and strong advantages among Russian Jewish youth, that could not be accounted for by differences in socioeconomic backgrounds. His findings, therefore, suggest the salience of both structural and cultural factors.

Importantly, Perlmann's results also suggest the salience of political, social, and economic power and participation for a group's aggregate school progress. Among the Irish, for example, these factors appear to account for the convergence of Irish and Yankee high school entry and school enrollment rates in Providence between 1880 and 1890. These kinds of factors may be the source of the city-to-city variations in progress differentials between immigrants as a whole and children of native whites reported by Olneck and Lazerson; variations among cities in rates of parochial versus public school attendance; and the varying patterns of the effects of neighborhood ethnic composition on the school attendance rates of ethnic groups in mid-nineteenth century Chicago and elsewhere reported by Galenson.[78] Findings such as these suggest the importance of placing the results for individuals within the contextual dynamics affecting the choices and opportunities faced by immigrant communities.

In contrast to Perlmann's findings—and those reported by Olneck and Lazerson—Jacobs and Greene concluded from their analyses of individual-level data from the 1910 decennial census, that controls for parental occupation, self-employment, recency of immigration, region, urbanicity, parental literacy, parental English ability, and presence of the father in the household generally account for observed variations in school enrollment among ethnic groups, and between the offspring of immigrants and those of native white parents.[79] They claim that remaining differentials are best explained by variations in economic opportunity. Nevertheless, even in Jacob and Greene's data, disparities in school enrollment favoring Jewish immigrant children over the children of Italians and Poles are quite robust.

But Stephen Steinberg, Selma Berrol, and John Bodnar—like Colin Greer—argue that only once they secured a favorable economic foothold through means other than schooling, were Jews able to act upon values conducive to lengthier schooling.[80] If Steinberg, Bodnar, and Berrol are correct, differentials in educational attainment favoring the children of Jewish immigrants should occur only above certain socioeconomic thresholds. We might also expect that the average socioeconomic level of a group, creating a socioeconomic context or environment beyond that of the family, to explain—to some degree—variations among ethnic groups in education attainment. No such patterns are evident in Perlmann's or other quantitative data. Moreover, the atypical use of the schools by Jewish immigrants does not mean that Jews relied solely upon schooling as a means of attaining social mobility. Certainly, as Selma Berrol and others emphasize, Jews, like other immigrants, relied as well on commercial and manufacturing avenues that did not require lengthy schooling.[81] Commercial avenues were particularly favorable for Jews.[82] Still, it is quite evident from later Census Bureau data that, at the least, Jewish men attained considerably more schooling than would be expected based upon their socioeconomic origins, and further, that the unusual occupational mobility they subsequently experienced is very largely attributable to their atypically lengthier schooling. Other groups, as well, including the Irish, experienced occupational mobility during the first half of the twentieth century that can be largely attributed to education attainment higher than expected on the basis of social origins.[83] Indeed, Perlmann found that in Providence, by 1925, "upward mobility among the sons of immigrant workers was transformed from something rarely accomplished with the aid of secondary schooling to something rarely accomplished without it."[84] Thus, at least for certain groups, the "legend" of immigrants availing themselves of American schools as a path to individual success holds considerable truth.

Those challenging "cultural" explanations for variations in ethnic group education success appear to believe that the proponents of cultural explanations assume "culture" to be "primordial" and static values that are impervious to constraints and opportunities, and are independent of skills, resources, and objective historical circumstances.[85] In fact, some proponents of the salience of culture for explaining education outcomes have taken care to qualify their arguments by noting that culture is not free-floating, nor does it exist in a vacuum, impervious to material and historical circumstances.[86] Nevertheless, the fact that multivariate analyses of individual-level data embody a logic by which the effects of variables are partitioned, isolated, presumed "independent" of one another, and compared to determine "relative" strengths, leads to the facile identification of certain variables with objective "structural" influences, and others, among them ethnic group membership, with stable "cultural" influences. The problem with this formulation is that "structural" variables such as "social class," and "cultural" variables such as "ethnicity" are not—as David Hogan has observed—"entities or nominalistic clusters of discrete characteristics that can be measured for their relative causal power in determining educational behavior as ethnocultural historians attempt."[87] Rather, class and ethnicity are cultural processes or forms of experience shaped by the interaction between the structure of class relations and cultural traditions.

Recent work in this area is much more self-conscious than previous work about the complexities of the meaning of, and relationships between, "culture" and "structure." Rather than culture being viewed as autonomous, it is viewed as "socially embedded, [and] related to present or previous social, political and economic relations,"[88] and is understood as partly "a product of factors such as a previous location in a specific class system, state system or natural environment, or rather all of these combined."[89] Recent work also highlights the significance of immigrant community contexts for not only objective opportunities, but also for "the attitudes, values and outlooks common in people from different groups."[90] Hans Vermeulen and Tijno Venema illustrate this approach in their comparison of the educational and occupational mobility experienced by Italian and Greek immigrants in the first half of the twentieth century.[91] They show that despite comparable peasant background, family social structure, cultural codes, urbanicity, initial occupations in the United States, gender roles, and opportunity structures, Greeks were quicker to attain middle-class status, and their offspring enjoyed atypically high education attainment. They attribute these differentials to the effects of stronger contacts between peasants and the outside world in Greece than in Italy, more equal patterns of land distribution in Greece than in Italy, higher levels of schooling and literacy among peasants in Greece as compared with those in Italy, significant levels of ethnic employment among Greek immigrants in the United States facilitated by the success of initially small-scale Greek immigrant entrepreneurs, and immigrant community institution-building and cohesion. "Culture" and "structure" are compounded here, and community context is highly salient. In contemporary sociological terminology, Greek immigrants developed communities stronger in "social capital"[92] than did Italian immigrants, and enjoyed the benefits of this circumstance.

A more nuanced understanding of culture and structure finds expression in the analysis of families' and communities' education strategies as one aspect of dynamically sustaining and refashioning—in shifting contexts of a technologically changing, wage-labor, capitalist system of production—what may be thought of as a "family economy."[93] Immigrant families required sustenance, and children and adults each contributed to that sustenance. The requirements that compulsory education imposed, and the possibilities that post-compulsory schooling offered, were evaluated in the context of what was required for the economic sustenance of the family, though the strategies deemed appropriate, and the trade-offs that were acceptable varied culturally within and among ethnic groups. The concepts of education strategies and family economy help to explain behavior with respect to schooling at any one time, and, more importantly, help to account in shifts in behavior respecting schooling, and to explain a trend toward convergence across ethnic groups in education attainment—during the second quarter of the twentieth century—that would have been unanticipated on the basis of analyses stressing only "values" or cultural preferences. This trend cannot be unambiguously attributed to changes in values or to changes in structure because of the inseparability of the ideology of family life and its material base.[94]

Miriam Cohen, for example, has emphasized the pragmatic adaptation of Southern Italian immigrants to increasingly limited opportunities for youth labor and the need for further schooling as a prerequisite to economic security during the 1930s and afterward, resulting in patterns of school attendance and attainment more characteristic of the American norm.[95] These shifts, according to Cohen, "occurred not because Italians adopted middle-class values of individualism . . . Italian families still depended on the contributions of all family members. However, important changes in the demographic structure of the Italian community, and in the employment structure of New York affected the educational patterns of Italian-Americans children, particularly of daughters."[96] In particular, Cohen notes that shifting patterns of opportunity in office work, requiring further schooling, resulted in a reversal of the "traditional" pattern of Italian boys receiving more schooling than their sisters. Stephen Lassonde—noting the attraction to vocationally relevant curricula, especially the commercial track—concurs with Cohen that continuation in high school among the children of Southern Italian immigrants represented less the adoption of "American" achievement norms and middle-class aspirations, as it did adapting lengthier schooling as a means to satisfy older, more established aspirations. David Hogan makes a similar argument regarding the trajectory of educational attainment among Chicago's Slavic communities.[97]

CONCLUSION

Historical scholarship on European immigrants and American public schools has, then, since the mid-1970s, helped to repudiate early revisionists' simplistic inversions of Cubberlyian[98] or Creminesque accounts of American education. Most significantly, it has incorporated the collective, familial, and individual perspectives and actions of immigrants themselves, and has demonstrated that immigrant *appropriation* of schooling is a more valid interpretative lens for comprehending the education of immigrants in American schools than are elite domination or social control. It has demonstrated the salience of broader trends in norms and cultural representations of childhood, youth, and adulthood for the education of immigrants. It has shown how immigrant groups, particularly the second generation, utilized the schools to reorganize themselves as recognized and respected American ethnic groups. It has also shown, however, that the political success of immigrants as "ethnics" was the obverse of the failure of working people to solidify as a working class. Finally, it has furthered the efforts of historians and social scientists interested in the past to rethink in more dynamic ways the nexus between culture and structure in accounting for patterns of ethnic achievement and attainment in the schools. In its greater complexity and sophistication, post-revisionist scholarship on European immigrants and American pubic schools is a worthy body of work to review in a volume honoring Carl F. Kaestle, whose own work has done so much to illuminate the complexities of an earlier century.

Coda I: A Note on the Historiography of Non-European Immigrants and Schooling

The historiography of immigrants and American schooling has been dominated by the study of European immigrant groups.[99] Indeed, the "immigrant story" that has been the dominant template for the "American story" is the story of European immigrants.[100] Yet, between 1870 and 1882, the year of the passage of the Chinese Exclusion Act, over 190,000 Chinese entered the United States. Even between 1883 and 1924, over 87,000 Chinese entered the country. Between 1890 and 1924, almost 360,000 Japanese entered the United States. Finally, between 1900 and 1930, almost 730,000 Mexicans entered the United States, though large numbers of these returned to Mexico in subsequent deportations.[101] The connotation of pre-1965 immigration as "European" thus omits from the symbolic construction of the generic immigrant important elements, unsurprisingly those who could not be imagined as "becoming white."[102]

The few histories of the education of Mexican immigrants in the United States in the first half of the twentieth century concentrate on the struggles by Mexican American communities and developing political organizations against the official and unofficial segregation of the children of Mexicans into separate and highly unequal schools.[103] The schools provided for children of Mexican immigrants in the first several decades of the twentieth century, in states like Texas, were poorly funded, oriented toward non-academic vocational preparation, and characterized by cultural and linguistic suppression and the abuse of standardized mental testing. While school walkouts and boycotts protesting segregation occurred as early as

1910 in San Angelo, Texas, segregation and inferior schooling for Mexican Americans became entrenched in the 1930s, and despite later legal victories against segregationist policies and practices, persisted into the 1960s. Further, despite challenges in the 1920s to legal proscriptions against Spanish in Texas schools, "no-Spanish" policies also persisted into the 1960s.

Historians have also paid a modicum of attention to the contradictory efforts to Americanize Mexican immigrants and their children.[104] Americanization of Mexican immigrants aimed to assimilate Mexicans to the use of English and the emulation of American ways of living, while at the same time—through segregation and subordination—marking and preserving the "racial frontier" that placed them, and, especially, Asians beyond the boundaries of full citizenship.[105]

It is worth noting that with the aid of the Mexican government, efforts were made in California during the 1920s to establish afterschool "Mexican schools" patterned after Hebrew and Japanese schools, where pupils could study Spanish, Mexican geography history, and native arts, though the extent to which these schools were patronized is unknown.[106] However, the detailed study of the pre-1940 twentieth-century Mexican immigrants' experiences with American schooling, comparable to the work on eastern European immigrants in northern and eastern cities, remains a task for the future, one which will very likely be undertaken by the increasing number of Chicano educational historians.[107]

Similarly, the historiography of the schooling of Asian immigrants in the early decades of the twentieth century concentrates on resistance to segregation and racism,[108] and on native whites' efforts to "Americanize" Asian immigrants, particularly on efforts to suppress Japanese American supplementary language schools.[109] Reed Ueda has, however, called attention to the manner in which Japanese American students in Hawaii capitalized upon the opportunities afforded by public secondary schooling.[110]

Japanese immigrants in Hawaii first established native language schools in the early 1890s. Initially, on the assumption that Japanese workers and their families would eventually return to Japan, the schools offered as close a replica as possible of schooling in Japan. Over time—both in response to pressure by native whites to demonstrate that their schools were not antagonistic to American principles, and in recognition of the increasingly attenuated ties between

second-generation Japanese Americans to their parents' homeland—
these schools increasingly limited the scope of their curriculum,
replacing Japanese materials with materials written in the United
States, and confining instruction to language.[111] In these tendencies,
the Japanese language schools paralleled other ethnically-based
schools, such as Polish Catholic schools in Chicago and elsewhere.
Like ethnic language schools more generally, Japanese language
schools were unable to stem the shift of the second-generation away
from their parents' language.[112] The schools are most interesting for
their role in communal politics, as Buddhists, Christians, and secular
Japanese vied to construct Japanese American identity, and for their
role in broader nativist attacks on the Japanese presence in Hawaii
and on the West Coast.[113]

While first-generation Japanese parents fretted that their children
would lose their language and their ties to Japanese culture—and
while nativists feared that the Nisei would "Japanize" the United
States—Japanese American second-generation youth were fashion-
ing ethnic identities and ethnic affiliations that would be the bases
for social mobility and for their relationship to American society
(insofar as they were permitted to exercise "ethinic options"). In an
analysis reminiscent of Paula Fass' account of European American
second-generation youths in New York City appropriating officially
sanctioned school activities to their own status and identity needs,
Reed Ueda has described how William McKinley High in Honolulu
in the 1920s and early 1930s was a site for the predominantly Nisei
student body for "the engagement of the micro-world of ethnic sub-
cultural identity and adolescent personality with the public world of
official, nationalist identity."[114] There—particularly through the
medium of the student newspaper, the *Pinion*—Japanese American
high school students advocated, and strove to adopt, Standard English
and to abandon pidgin, while at the same time they appropriated
American "vocabularies of public life,"[115] associating democracy and
equal rights with Americanism to elaborate a pluralistic ethnic identity.

While we are accustomed to associating Mexican and Asian immigrants with post-1965 immigration, the substantial presence of both
Mexicans and Asians in the United States during the same period as
mass European immigration offers an opportunity for comparative
analysis that would yield a much fuller understanding of the dynamics of immigrant acculturation and incorporation within the thoroughly racialized American social order than has been accomplished

to date. It is to be hoped that historians of education of the future will participate in conducting such inquiries.

Coda II: Immigrants Then, Immigrants Now

The United States has, since passage of the 1965 Immigrant Act, once again experienced mass immigration. Perceptions and expectations of today's "new immigrants" are shaped by (not always accurate) perceptions of how immigrants in the past fared, and were incorporated into American society.[116] A large and growing social scientific literature examines the social, economic, and educational experiences and trajectories of contemporary immigrants and their offspring.[117] In some cases, social scientifically-oriented historians have undertaken comparisons of contemporary immigrants with immigrants of the pre-World War II twentieth century.[118] In addition to the empirical findings concerning recent immigrants, and the comparison of these to findings about previous immigrant groups, contemporary scholarship has introduced concepts and lines of analysis, that if applied retrospectively to the immigrant populations of the past, might prove—and in a few instances have already proven—illuminating. These include most prominently *segmented assimilation*[119] and *social capital*. [120]

Segmented Assimilation

"Segmented assimilation" refers to distinctive trajectories of immigrant incorporation culminating in highly divergent socioeconomic destinations. Some groups are assumed to experience the kind of smooth intergenerational social mobility, including progressive acculturation and increased education attainment, that is associated with images of "straight-line assimilation." Such straight-line assimilation is often assumed by adherents of the segmented assimilation model to have characterized the European immigrants of the early twentieth century. Other groups—particularly those whose skin is darker—are presumed to undergo processes of "racialization" that identify them with African Americans as subordinated minorities, confine them to impoverished inner-city residence, along with the ineffective schools found in such locations, subject them to encounters with pervasive discrimination, and, in the second-generation, place them at risk of adopting "oppositional" or "resistant" identities and practices associated with native minorities, thereby producing socioeconomic stagnation or decline. Still other groups capitalize on ethnic identity and solidarity, the economic opportunities provided by ethnic niches or enclaves, and ethnically sanctioned, community-enforced norms to promote social mobility in the form of "accommodation without assimilation."[121] Which trajectory particular groups experience is attributed to the interplay of the human capital of the immigrant generation, contexts of reception that range from hostility to programs of welcome and assistance, the nature of pre-existing communities of co-ethnics, and bifurcation of the labor market into low-paying, "dead-end" low-skill jobs and high-paying, stable, skilled occupations requiring lengthy schooling.

The applicability of the segmented assimilation model as a contrast between the processes of incorporation and subsequent trajectories characterizing contemporary immigrants—who are disproportionately east and south Asian, and Latin American, especially Mexican—and the processes characterizing the incorporation and trajectories experienced by the southern and eastern European immigrants of the early twentieth century is intensely debated. Some scholars are convinced that segmented assimilation, and the "second-generation decline"[122] that it implies for some groups, is an indisputable fact.[123] Other scholars believe that it is premature to extrapolate from the relatively young members of contemporary second-generations to conclusions about the ultimate intergenerational prospects of particular groups. [124]

Recent empirical analyses suggest that both socioeconomic resources and ethnic group membership are salient in promoting or constraining the educational opportunities of contemporary immigrants. In one important data set from south Florida, second-generation Haitians and West Indians who were approximately twenty-four years old in 2002, completed *more* schooling than would have been expected on the basis of their socioeconomic origins.[125] Evidence of the possible effects of racial discrimination resulting in lower incomes is, however, present in these data. Analyses of 1998 and 2000 Current Population Survey data show that second- and third-generation immigrant populations of all kinds exceed earlier first-generation populations in education attainment, occupational status, and incomes, though progress for Mexican, Puerto

Rican, and Central American second and third generations is notice-ably slower than that of other groups, and may stall in the third generation.[126] In San Diego in 2001–2003, while Vietnamese second-generation youth had acquired disproportionately lengthy schooling, Mexican second-generation youth were disproportionately found among those with low education attainment.[127] Most disturbing about this result is that over 70 percent of the Mexican education disadvantage persists even when parental socioeconomic status, family size, presence of two adults, and parental nativity are held constant. There is also evidence that many second-generation immigrants *are* perceiving the American opportunity structure in racially hierarchical terms, and are locating themselves as racially disfavored, generating, in turn, a decline in "immigrant optimism."[128]

Direct comparison between the trajectories of contemporary immigrants and those of the past has been attempted most notably by Joel Perlmann.[129] His findings, comparing the progeny of non-Jewish, south-central European immigrants of the early twentieth century with the progeny of contemporary Mexicans who immigrated into the United States, suggest that education mobility in relation to the attainment of native whites is considerably less for Mexican males, but quite comparable for Mexican women. While Perlmann does not find evidence for the strongest versions of the segmented assimilation model, his results concerning the disproportionately high high-school dropout rates for the offspring of Mexican immigrants—when coupled with the increasingly severe economic penalties for low levels of education compared with those faced by pre-World War II immigrant offspring—are cause for alarm. So, too, is the overall increase in levels of income inequality that distinguishes the contemporary context from the contexts in which prior generations of immigrants lived.

Assertion of segmented assimilation as characteristic of contemporary immigrant incorporation *in contrast* with assimilation patterns of the past has prompted some scholars to revisit assumptions about uniformly positive trajectories for European immigrant groups in the pre-World War II period. As explained earlier, the trajectory of education attainments of various immigrant groups' offspring varied. But whether the attainments of any second-generation groups *fell*—either in relation to the attainments of their parents' generation or in relation to similarly aged native-born offspring of native-born parents—as the segmented assimilation model would anticipate, is a distinct question. While this question has not been examined in great detail, Sharon Sassler has analyzed 1920 U.S. Census data with a view to attempting a preliminary answer.[130] While Sassler was confined to comparing distinct generations at a single point in time, rather than subsequent generations over time, and to analyzing school *enrollment* rather than eventual education attainment, her results are nevertheless interesting. Sassler found that for boys aged fifteen to eighteen years old, education improvement was evident across generations for all ethnic groups. However, when Sassler took into account the effects of household head's socioeconomic status—as well as other household resources and demographic characteristics—for the Irish and Poles, she found some modest evidence of *downward* trends in the net likelihoods of male school enrollment from first to second generation, though in the case of the Poles, the decline is statistically insignificant. Among girls, Sassler found little, if any intergenerational improvement, in the chances of school enrollment among Italians and Poles, and, after introducing controls for socioeconomic and demographic factors—she found statistically insignificant declines in enrollment likelihoods for Italians, Poles, and Jews. Perhaps over-interpreting her rather modest results, Sassler concluded that "disparities in schooling [across ethnic groups] remained beyond the second-generation . . . The assimilation process was clearly not completed by the third generation . . . segmented assimilation is not a recent development."[131]

Perhaps more compelling than Sassler's results are those of Alba, Lutz, and Vesselinov's analyses of 1986–1994 National Opinion Research Center (NORC) General Social Survey data which show that among European groups, there is no correspondence between the literacy rates of the immigrant first generation as of 1910 and the educational attainment of the subsequent third generation, and that the effects of variations in the wages of the occupations occupied by various immigrant groups in 1910 on the education attainment of later generations are evident only for the disproportionately high education attainments of Russian Jews.[132] These researchers conclude that "over the two generational transitions reflected here, assimilation has largely eradicated the disadvantages suffered by groups of peasant origins in Europe," thus offering no support for the application of the segmented assimilation model to pre-World War II European groups.[133] Alba, Lutz, and Vesselinov's results, however, are due in part to their exclusion of Mexicans from their analysis, in contrast to analyses of the same data by George Borjas.[134] Evidence of persistent education disadvantage among the descendants of early twentieth-century Mexican immigrants offers ironic parallels to the assertions by segmented assimilation adherents that the offspring of contemporary Mexican immigrants are distinctively disadvantaged.

A key idea of the segmented assimilation model is "racialization." The idea is that contemporary immigrants whose skin is darker than those of other immigrants (*e.g.*, Haitians, West Indians, Dominicans, Mexicans) will be incorporated into America's racial order on unfavorable terms. Certainly there are survey and ethnographic data supporting this possibility. In some contexts, immigrant newcomers to American schools are initiated into the terminology and social practices of racial classification, leaving them excluded or marginalized.[135] Over time, second-generation immigrants acquire racial identifications that are foreign to their parents.[136] Haitian and West Indian students in heavily black inner-city schools often adopt identities as "black," along with the cultural practices associated with this identity.[137]

Nevertheless, racial classifications are dynamic and fluid, and the consequences of racializing processes are not predictable. While from the perspective of the early twenty-first century, twentieth-century European immigrants may be perceived as having been readily assimilable because of their racial and cultural "similarity" to native-born white Americans,[138] such was not the perception at the time. As Matthew Frye Jacobson and David Roediger have shown, southern, central, and eastern European immigrant groups had to *become* "white" and "ethnic" through complex, difficult, and contested processes involving legal decisions, shifts in elite's cultural interpretations, labor organizing and mobilization, shop-floor job divisions and social relationships, urban politics, intra-Catholic, cross-national marriage, and residential patterning.[139] Participation of the immigrant second generation in World War II also contributed to inclusion within an encompassing, Caucasian citizenry.[140] At a more local level, we may suppose that while participation in school extracurricular activities was ethnically patterned, the relatively high mingling of second-generation European Americans compared to the disproportionate absence of African Americans from school clubs (or their disproportionate over-representation in sports like track) in New

York City's high schools in the 1930s and 1940s contributed to immigrants' children finding their way to the right side of the color line.[141]

Some scholars counter the anticipation of negative racialization of contemporary immigrants by assuming that the diversity of immigrants will render stereotyping more difficult and the negotiation of racial and ethnic boundaries more possible than in the past,[142] by suggesting that racial distinctions may be less rigid and more blurry than in the past, and that acceptance of diversity is increasingly normative,[143] and by calling attention to the increasing socioeconomic diversity among Asians *and* African Americans, increasing intermarriage among Asians and whites, and the possibly diminishing salience of even the white-black divide.[144] One has only to contrast the virulent and long-lasting racism directed at Asian Americans with the diminished salience of race now for, at the least, middle-class Asian Americans[145] to recognize the possibility of the "declining significance of race."[146] In some contexts, an ethnic continuum, not the black-white dichotomy, does promote hybrid groupings, and fluid exchanges across group boundaries, leading to shared local identification as *e.g.*, "New Yorker," if not as "Americans," which remains associated with being white.[147] On the other hand, awareness of the racial stigma attaching to African Americans motivates at least some immigrant adults and youths to work at distancing and distinguishing themselves from African Americans.[148] As in the past, the offspring of not-already-white immigrants may gain privilege and advantage at the expense of Americans whose ancestors were first brought to what is now the United States in 1619.

Social Capital

At the most general level, "social capital" refers to "the ability of actors to secure benefits by virtue of membership in social networks or other social structures"[149] Social capital functions to facilitate (1) the transmission of and adherence to shared norms, (2) the cultivation of trust, reciprocal assistance, and mutual support, and (3) the provision of institutional resources and opportunities necessary for success and social mobility. Social capital works to make values efficacious,[150] so it may well be differences in social capital rather than differences in values that account for different education outcomes among groups and individuals.[151] Sociologists have applied the concept of social capital to explain how the solidarity and "institutional completeness" of particular immigrant communities, especially the Vietnamese, have contributed to the academic success of immigrant and second-generation students in American schools,[152] and to explain how weakness of social capital both outside and within schools enjoyed by Mexican American students has contributed to their difficulties in attaining school success.[153] Where they do exist, networks among Mexican immigrant students of achievement-oriented peers facilitate the sharing of school-related information, material resources, and school-based knowledge, and the adoption of collective achievement strategies.[154]

Although Claudia Goldin and Lawrence F. Katz, Reed Ueda, and John Rury have each utilized the concept of social capital in historical analysis of American education, historians of immigrants and schooling have not utilized the concept to analyze variations in the educational trajectories of different ethnic groups.[155] The concept might well prove applicable in accounting for differences that have perhaps been too readily explained by distinctive cultural values regarding education. For example, Olneck and Lazerson advanced a strong culturalist explanation for the differences between the school achievement of the children of Eastern European Jewish immigrants and those of Southern Italian immigrants.[156] Perhaps Jewish immigrant communities were stronger in the kinds of social capital that have proven educationally efficacious for the most successful of contemporary immigrant groups. German Jewish communities engaged in substantial institutional building aimed at educating their co-religionists. The Educational Alliance stands is the most striking example.[157] The Jewish immigrant community had even engaged in efforts at communal governance through experimenting with a version of the European kehilla.[158] The Jewish immigrant community, as we have seen, was capable of rapid mobilization when it believed its educational interests to be threatened.[159] Jews also took advantage of institutional opportunities provided by native Protestants to construct mutually supportive intergenerational networks that facilitated "bridging" into the broader society. Boston's West End House stands as an important example of this.[160] According to Ueda, Russian Jews in Boston in the early 1900s, like Honolulu's Japanese community, represented a "highly solidaristic" immigrant community, characterized by "dense networks of cooperative and coordinated roles anchored in families and communal subgroups," which provided the core membership of West End House, and utilized it to capitalize on the community's "mobility ethic operationalized by intergenerational partnership" to "propel . . . a voluntary quest of educational opportunity."[161]

On the other hand, traditions of familialism in southern Italy did not predispose Italian immigrants to construct the kinds of social networks that the Jewish immigrant community appears to have constructed. One eminent historian of immigrant life described Southern Italian communities in the United States as evidencing "a marked incapacity . . . for organizational activity."[162] As noted, stronger social capital enabled Greek immigrants to achieve more commercial success than Italians, even when the two groups shared values. Closer examination by historians of immigrant communities and families—and of the strategies they and their children adopted in relation to schooling through the analytic prism of "social capital"—could well pay dividends in our understanding of educational achievement and attainment among diverse immigrant groups.

NOTES

1. Barbara Finkelstein, "Education Historians as Mythmakers," *Review of Research in Education* 18 (1992): 255-297.
2. Jerry Jacobs and Margaret Greene, "Race and Ethnicity, Social Class, and Schooling," in *After Ellis Island: Newcomers and Natives in the 1910 Census*, ed. Susan Cotts Watkins (New York: Russell Sage, 1994), 209–55.
3. John Bodnar, "Materialism and Morality: Slavic-American Immigrants and Education, 1890–1940," *Journal of Ethnic Studies* 3 (Winter 1976): 1.
4. Colin Greer, *The Great School Legend: A Revisionist Interpretation of American Public Education* (New York: Basic Books, 1972), 119.
5. Ibid., 26.
6. Lawrence A. Cremin, *The Transformation of the School: Progressivism in American Education, 1876–1957* (New York: Vintage Books, 1961).
7. Colin Greer, *The Great School Legend*, 80.
8. Joel H. Spring, *Education and the Rise of the Corporate State* (Boston: Beacon Press, 1972).
9. Michael B. Katz, "The Origins of Public Education: A Reassessment," *History of Education Quarterly* 16 (Winter 1976): 394.
10. Michael B. Katz, *Class, Bureaucracy, and Schools: The Illusion of Educational Change in America* (New York: Praeger, 1971), 39.
11. Samuel Bowles and Herbert Gintis, *Schooling in Capitalist America: Educational Reform and the Contradictions of Economic Life* (New York: Basic Books, 1976).
12. Marvin F. Lazerson, *Origins of the Urban School: Public Education in Massachusetts, 1870–1915* (Cambridge, MA: Harvard University Press, 1971).

13. David Tyack, *The One Best System: A History of American Urban Education* (Cambridge, MA: Harvard University Press, 1974).
14. Ibid., 232.
15. Ibid., 242.
16. Marvin F. Lazerson, "Revisionism and American Educational History," *Harvard Educational Review* 43 (Summer 1973): 269–83; Allan Stanley Horlick, "Radical School Legends," *History of Education Quarterly* 14 (Summer 1974): 252–58.
17. Carl F. Kaestle, "Social Reform and the Urban School," *History of Education Quarterly* 12 (Summer 1972): 211–28.
18. David John Hogan, *Class and Reform: School and Society in Chicago, 1880–1930* (Philadelphia: University of Pennsylvania Press, 1985); Stephen E. Brumberg, *Going to America Going to School: The Jewish Immigrant Public School Encounter in Turn-of-the-Century New York City* (New York: Praeger, 1986); Paula Fass, *Outside In: Minorities and the Transformation of American Education* (New York: Oxford University Press, 1989).
19. Stephen E. Brumberg, *Going to America*, 79.
20. Paula Fass, *Outside In*, 59.
21. Paula Fass, *Outside In*, passim.
22. Ronald D. Cohen and Raymond A. Mohl, *The Paradox of Progressive Education: The Gary Plan and Urban Schooling* (Port Washington, NY: Kennikat, 1979).
23. Stephen Lassonde, *Learning to Forget: Schooling and Family Life in New Haven's Working Class 1870–1940* (New Haven: Yale University Press, 2005); Paula Fass, "Immigration and Education in the United States," in *A Companion to American Immigration*, ed. Reed Ueda (Malden, MA: Blackwell, 2006), 492–512.
24. Stephen Lassonde, *Learning to Forget*, 26.
25. Paula Fass, "Immigration and Education," 496.
26. Stephen Lassonde, *Learning to Forget*, 3; Paula Fass, "Immigration and Education," 496.
27. Paula Fass, *Outside In*, 77.
28. Selma Cantor Berrol, *Immigrants at School: New York City, 1898–1914* (New York: Arno, 1978).
29. Ibid., 140.
30. Selma Cantor Berrol, *Immigrants at School*.
31. Stephen E. Brumberg, *Going to America*, 128.
32. Ibid., 73.
33. Barbara Finkelstein, "Education Historians," 275.
34. Ronald D. Cohen and Raymond A. Mohl, *The Paradox of Progressive Education*, 108.
35. Stephen E. Brumberg, *Going to America*.
36. Paula Fass, *Outside In*, 10.
37. Ronald D. Cohen and Raymond A. Mohl, *The Paradox of Progressive Education*, passim.
38. Stephen E. Brumberg, *Going to America*, 72.
39. Selma Berrol, "Public Schools and Immigrants: The New York City Experience," in *American Education and the European Immigrant, 1840-1940*, ed. Bernard J. Weiss (Urbana: University of Illinois Press, 1982), 30-43.
40. Michael R. Olneck and Marvin F. Lazerson, "Education," in *Harvard Encyclopedia of American Ethnic Groups*, ed. Stephan Thernstrom (Cambridge, MA: Belknap, 1980).
41. Ronald D. Cohen and Raymond A. Mohl, *The Paradox of Progressive Education*, passim.

42. Judith R. Raftery, *Land of Fair Promise: Politics and Reform in Los Angeles Schools, 1885–1941* (Stanford: Stanford University Press, 1992).

43. David Hogan, "Education and the Making of the Chicago Working Class, 1880–1930," *History of Education Quarterly* 18 (Autumn 1978): 227–70; David A. Gerber, "Language Maintenance, Ethnic Group Formation, and Public Schools: Changing Patterns of German Concern, Buffalo, 1837–184," *Journal of American Ethnic History* 4 (Fall, 1984): 31–61; Dorota Praszalowicz, "The Cultural Changes of Polish-American Parochial Schools in Milwaukee, 1866–1988," *Journal of American Ethnic History* 13 (June 1994): 23–45.

44. Judith R. Raftery, *Land of Fair Promise*, passim.

45. Steven L. Schlossman, "Is There an American Tradition of Bilingual Education? German in the Public Elementary Schools, 1840–1919," *American Journal of Education* 91 (February 1983): 139–86.

46. Dorota Praszalowicz, "The Cultural Changes"; Jonathan Zimmerman, "Ethnics Against Ethnicity: European Immigrants and Foreign-Language Instruction, 1890–1940," *Journal of American History* 88 (March 2002): 1383–1404.

47. Leonard Covello, *The Heart is the Teacher* (New York: McGraw Hill, 1958); Deborah Dash Moore, *At Home in America: Second Generation New York Jews* (New York: Columbia University Press, 1981); David A. Gerber, "Language Maintenance, Ethnic Group Formation, and Public Schools: Changing Patterns of German Concern, Buffalo, 1837–184," *Journal of American Ethnic History* 4 (Fall, 1984): 31–61; Michael R. Olneck, "What Have Immigrants Wanted from Americans Schools? What Do They Want Now?" (keynote address, annual meeting off the Midwest History of Education Society, Chicago, IL, October 27, 2006).

48. Stephen E. Brumberg, *Going to America*, passim.

49. Raymond A. Mohl, "Schools, Politics, and Riots: The Gary Plan in New York City, 1914–1917," *Paedagogica Historica* 15 (June 1975): 39–72.

50. Maxine S. Seller, *Ethnic Communities and Education in Buffalo, New York: Politics, Power, and Group Identity, 1838–1978*. Community Studies Graduate Group, Paper no.1, (1979), State University of New York at Buffalo, Buffalo, NY, 1-136.

51. James W. Sanders, *The Education of an Urban Minority: Catholics in Chicago, 1833–1965* (New York: Oxford University Press, 1977); JoEllen McNergney, *For Faith and Fortune: The Education of Catholic Immigrants in Detroit, 1805–1925* (Urbana: University of Illinois Press, 1998).

52. John Ralph and Richard Rubinson, "Immigration and the Expansion of Schooling in the United States, 1890–1970," *American Sociological Review* 45 (December 1980): 943–54.

53. David W. Galenson, "Ethnic Differences in Neighborhood Effects on the School Attendance of Boys in Early Chicago," *History of Education Quarterly* 38 (Spring 1998): 17–35.

54. John Bodnar, "Materialism and Morality: Slavic-American Immigrants and Education, 1890-1940," *Journal of Ethnic Studies* 3 (Winter 1976): 1–19; David Hogan, "Education and the Making of the Chicago Working Class"; Ronald D. Cohen and Raymond A. Mohl, *The Paradox of Progressive Education.*

55. David A. Gerber, "Language Maintenance"; Michael R. Olneck, "What Have Immigrants Wanted."

56. Marvin Lazerson, "Understanding American Catholic Educational History," *History of Education Quarterly* 17 (Autumn, 1977): 297–317; James W. Sanders, *The Education of an Urban Minority*; Dorota Praszalowicz, "The Cultural Changes"; William J. Galush, "What Should Janek Learn? Staffing and Curriculum in Polish-American

Parochial Schools, 1870–1940," *History of Education Quarterly* 40 (Winter 2000): 395–417.
57. Stephen Lassonde, *Learning to Forget*, passim.
58. Ibid.
59. Paula Fass, *Outside In*; Stephen Lassonde, *Learning to Forget*.
60. Selma Canto Berrol, *Immigrants at School*; Maxine S. Seller, *Ethnic Communities and Education in Buffalo*; Stephen Lassonde, *Learning to Forget*.
61. Richard Rubinson, "Class Formation, Politics, and Institutions: Schooling in the United States," *American Journal of Sociology* 92 (November 1986): 519–48.
62. Stephen Lassonde, *Learning to Forget*, passim.
63. Michael R. Olneck, "Immigrants and Education in the United States," in *Handbook of Research on Multicultural Education*, 2nd ed. James A. and Louise Cherry Banks (San Francisco: Jossey-Bass, 2004), 391–92.
64. Paula Fass, *Outside In*, 109.
65. Herbert Gans, "Toward a Reconciliation of 'Assimilation' and 'Pluralism': The Interplay of Acculturation and Ethnic Retention," *International Migration Review* 31 (Winter 1997): 875–92.
66. Steven L. Schlossman, "Is There an American Tradition"; David A. Gerber, "Language Maintenance."
67. Deborah Dash Moore, *At Home*; Leonard Covello, *The Heart Is*; Jonathan Zimmerman, "Ethnics Against Ethnicity."
68. Jonathan Zimmerman, "Ethinics Against Ethnicity."
69. Michael R. Olneck, "What Have Immigrants Wanted."
70. Paul E. Peterson, *The Politics of School Reform 1870–1940* (Chicago: University of Chicago Press, 1985).
71. Michael R. Olneck, "What Have Immigrants Wanted."
72. Ira Katznelson, Kathleen Gile, and Margaret Weir, "Public Schooling and Working-Class Formation: The Case of the United States," *American Journal of Education* 90 (February 1982): 111–43; Ira Katznelson and Margaret Weir, *Schooling for All: Class, Race, and the Decline of the Democratic Ideal* (New York: Basic Books, 1985); Richard Rubinson, "Class Formation, Politics, and Institutions."
73. Ira Katznelson and Margaret Weir, *Schooling for All*, 92.
74. Samuel Bowles and Herbert Gintis, *Schooling in Capitalist America*; Richard Rubinson, "Class Formation, Politics, and Institutions," 532.
75. David K. Cohen, "Immigrants and the Schools," *Review of Educational Research* 40 (February 1970): 13–27; Michael R. Olneck and Marvin F. Lazerson, "The School Achievement of Immigrant Children: 1900–1930," *History of Education Quarterly* 14 (Winter 1974): 453–82; David Tyack, *One Best System*, passim.
76. Joel Perlmann, *Ethnic Differences: Schooling and Social Structure Among the Irish, Italians, Jews, and Blacks in an American City, 1880–1935* (Cambridge, MA: Cambridge University Press, 1988); Jerry Jacobs and Margaret Greene, "Race and Ethnicity."
77. Joel Perlmann, *Ethnic Differences*, passim.
78. David W. Galenson, "Neighborhood Effects on the School Attendance of Irish Immigrants' Sons in Boston and Chicago in 1860," *American Journal of Education* 105 (May 1997): 261–93; David W. Galenson, "Ethnic Differences in Neighborhood Effects."
79. Jerry Jacobs and Margaret Greene, "Race and Ethnicity."
80. Stephen Steinberg, *The Ethnic Myth: Race, Ethnicity, and Class in America* (New York: Atheneum, 1981); Selma Berrol, "Public Schools and Immigrants: The New York City Experience," in *American Education and the European Immigrant,*

1840–1940, ed. Bernard J. Weiss (Urbana: University of Illinois Press, 1982), 30–43; John Bodnar, *The Transplanted: A History of Immigrants in Urban America* (Bloomington: Indiana University Press, 1985); Colin Greer, *Great School Legend.*

81. Selma Berrol, "The Open City: Jews, Jobs, and Schools in New York City, 1880–1915," in *Educating an Urban People: The New York City Experience*, ed. Diane Ravitch and Ronald Goodenow (New York: Teachers College Press, 1981), 101–15; Selma Berrol, "Public Schools and Immigrants."

82. Joel Perlmann, "What the Jews Brought. East-European Jewish Immigration to the United States, *c.* 1900," in *Immigrants, Schooling and Social Mobility: Does Culture Make a Difference?* ed. Hans Vermeulen and Joel Perlmann (New York: St. Martins Press, 2000), 103–23.

83. Beverly Duncan and Otis Dudley Duncan, "Minorities and the Process of Stratification," *American Sociological Review* 33 (June 1968): 356–64.

84. Joel Perlmann, *Ethnic Differences*, 41.

85. Stephen Steinberg, *The Ethnic Myth*; Stephen Steinberg, "The Cultural Fallacy in Studies of Racial and Ethnic Mobility," in *Immigrants, Schooling and Social Mobility*, 61–71; Miriam Cohen, *Workshop to Office: Two Generations of Italian Women in New York City, 1900–1950* (Ithaca: Cornell University Press, 1992); Jerry Jacobs and Margaret Greene, "Race and Ethnicity."

86. Michael R. Olneck and Marvin F. Lazerson, "Immigrant School Achievement."

87. David Hogan, "The Making," 261.

88. Hans Vermeulen, "Introduction: The Role of Culture in Explanations of Social Mobility," in *Immigrants, Schooling and Social Mobility*, 13.

89. Ibid., 15.

90. Joel Perlmann, "Introduction: The Persistence of Culture versus Structure in Recent Work. The Case of Modes of Incorporation," *Immigrants, Schooling and Social Mobility*, 25.

91. Hans Vermeulen and Tijno Venema, "Peasantry and Trading Diaspora. Differential Social Mobility of Italians and Greeks in the United States," in *Immigrants, Schooling and Social Mobility*, 124–49.

92. Alejandro Portes, "Social Capital: Its Origins and Applications in Modern Sociology," in *Annual Review of Sociology* 24, ed. John Hagan (Palo Alto, CA: Annual Reviews, 1998), 1–24.

93. David Hogan, "Education and the Making"; John Bodnar, *The Transplanted*; Joel Perlmann, *Ethnic Differences*; Stephen Lassonde, *Learning to Forget.*

94. Stephen Lassonde, *Learning to Forget*, passim.

95. Miriam Cohen, "Changing Education Strategies Among Immigrant Generations: New York Italians in Comparative Perspective," *Journal of Social History* 15 (Spring 1982): 443–62; Cohen, *Workshop to Office*, passim.

96. Miriam Cohen, "Changing Education Strategies," 451.

97. Stephen Lassonde, *Learning to Forget*; David Hogan, "Educaton and The Making"; Hogan, *Class and Reform.*

98. Ellwood Cubberley, Professor of Education at Stanford University between 1905 and 1933, and author of the highly influential *Public Education in the United States* (Boston: Houghton Mifflin, 1919), interpreted the development of American education as a triumph of the democratic character of American society. See Lawrence A. Cremin, *The Wonderful World of Ellwood Patterson Cubberley: An Essay on the Historiography of American Education* (New York: Bureau of Publications, Teachers College, Columbia University, 1965). Dates of Cubberly's appointment at Stanford are taken from Education Encyclopedia, "Ellwood Cubberley (1868–1941), Education

and Career, Contribution," StateUniversity.com, http://education.stateuniversity
.com/pages/1893/Cubberley-Ellwood-1868-1941.html.
99. Eileen H. Tamura, "Asian Americans in the History of Education: An Historiographical Essay," *History of Education Quarterly* 41 (Spring 2001): 8–71.
100. Silvia Wynter, *Do Not Call Us Negros: How "Multicultural" Textbooks Perpetuate
Racism* (San Francisco: Aspire, 1990).
101. Susan B. Carter et al., *Historical Statistics of the United States*, Millennial Edition On
Line (Cambridge University Press, 2006), Tables Ad136–148, Ad162–172,
http://hsus.cambridge.org.ezproxy.library.wisc.edu/HSUSWeb/toc/tableToc.do?id
=Ad136–148, and http://hsus.cambridge.org.ezproxy.library.wisc.edu/HSUSWeb/
table/showtablepdf.do?id=Ad162–172; Joel Perlmann, *Italians Then, Mexicans Now:
Immigrant Origins and Second-Generation Progress, 1890–2000* (New York: Russell
Sage Foundation, 2005).
102. Michael R. Olneck, "Immigrants and Education."
103. Mario T. Garcia, *Mexican Americans: Leadership, Ideology, & Identity, 1930–1960*
(New Haven: Yale University Press, 1989); Gilber G. Gonzalez, *Chicano Education
in the Era of Segregation* (Philadelphia: Balch Institute Press, 1990); Guadalupe San
Miguel, Jr., "The Struggle against Separate and Unequal Schools: Middle Class Mex-
ican Americans and the Desegregation Campaign in Texas, 1929–1957," *History of
Education Quarterly* 23 (Autumn, 1983): 343–59; Guadalupe San Miguel, Jr., *"Let
All of Them Take Heed": Mexican Americans and the Campaign for Educational
Equality in Texas* (Austin: University of Texas Press, 1987); Guadalupe San Miguel,
Jr. and Richard R. Valencia, "From the Treaty of Guadalupe Hidalgo to *Hopwood*:
The Educational Plight and Struggle of Mexican Americans in the Southwest," *Harvard Educational Review* 68 (Fall 1998): 353–412.
104. Mario T. Garcia, "Americanization and the Mexican Immigrant, 1880–1930," *Journal of Ethnic Studies* 6 (Summer, 1978): 19–34.
105. Frank Van Nuys, *Americanizing the West: Race, Immigrants, and Citizenship,
1890–1930* (Lawrence: University of Kansas Press, 2002).
106. George J. Sanchez, *Becoming Mexican American: Ethnicity, Culture and Identity in
Chicano Los Angeles, 1900–1945* (New York: Oxford University Press, 1993).
107. Ruben Donato and Marvin Lazerson, "New Directions in American Educational History: Problems and Prospects," *Educational Researcher* 29 (November 2000): 4–15.
108. Charles Wollenberg, *All Deliberate Speed: Segregation and Exclusion in California
Schools, 1877–1975* (Berkeley: University of California Press, 1976); David K. Woo,
*Growing Up Nisei: Race, Generation, and Culture Among Japanese Americans of Cal-
ifornia, 1924–49* (Urbana: University of Illinois Press, 2000).
109. Eileen H. Tamura, "The English-Only Effort, the Anti-Japanese Campaign, and Language Acquisition in the Education of Japanese Americans in Hawaii, 1915–1940,"
History of Education Quarterly 33 (Spring, 1993): 37–58; Eileen H. Tamuara, *Amer-
icanization, Acculturation, and Ethnic Identity: The Nisei Generation in Hawaii*
(Urbana: University of Illinois Press, 1994); Noriko Asato, "Mandating Americaniza-
tion: Japanese Language Schools and the Federal Survey of Education in Hawaii,"
History of Education Quarterly 43 (March 2003): 10–38; Noriko Asato, "Ousting
Japanese Language Schools: Americanization and Cultural Maintenance in Washing-
ton State, 1919–1927," *Pacific Northwest Quarterly* 94 (Summer 2003): 140–50;
Norkio Asato, *Teaching Mikadoism: The Attack on Japanese Language Schools in
Hawaii, California, and Washington, 1919–1927* (Honolulu: University of Hawaii
Press, 2006).

138   MICHAEL R. OLNECK

110. Reed Ueda, "Second-Generation Civic America: Education, Citizenship, and the Children of Immigrants," *Journal of Interdisciplinary History* 24 (Spring, 1999): 661–81.
111. Noriko Asato, *Teaching Mikadoism*, passim.
112. Joshua A. Fishman, "Ethnic Community Mother Tongue Schools in the U.S.A.: Dynamics and Distributions," *International Migration Review* 14 (Summer 1980): 235–47; Eileen H. Tamura, "The English-Only Effort"; Noriko Asato, *Teaching Mikadoism.*
113. Eileen H. Tamura, "The English-Only Effort"; Noriko Asato, *Teaching Mikadosim.*
114. Reed Ueda, "Second-Generation Civic America: Education, Citizenship, and the Children of Immigrants." *Journal of Interdisciplinary History* 24 (Spring 1999), 664; Mary C. Waters, *Ethnic Options: Choosing Identities in America* (Berkeley: University of California Press, 1990)
115. Robert Wuthnow, ed., *Vocabularies of Public Life: Empirical Essays in Symbolic Structure* (Routledge: London and New York, 1992).
116. Martha Montero-Sieburth and Mark La Celle-Peterson, "Immigration and Schooling: An Ethnohistorical Account of Policy and Family Perspectives in an Urban Community," *Anthropology & Education Quarterly* 22 (December 1991): 300–25; Ellen Bigler, "Telling Stories: On Ethnicity, Exclusion, and Education in Upstate New York," *Anthropology & Education Quarterly* 27 (June 1997): 186–203.
117. *E.g.*, Alejandro Portes and Rubén Rumbaut, *Legacies: The Story of the Immigrant Second Generation* (Berkeley: University of California Press, 2001). See Michael R. Olneck, "Immigrants and Education" for a review and synthesis of this literature.
118. Joel Perlmann and Roger Waldinger, "Second Generation Decline? Children of Immigrants, Past and Present—A Reconsideration," *International Migration Review* 31 (Winter 1997): 893–922; Joel Perlman, *Italians Then.*
119. See Alejandro Portes and Min Zhou, "The New Second Generation: Segmented Assimilation and Its Variants," *Annals of the American Academy of Political and Social Science* 530 (November, 1993): 74–96; Alejandro Portes, "Segmented Assimilation Among New Immigrant Youth: A Conceptual Framework," in *California's Immigrant Children: Theory, Research and Implications for Educational Policy*, ed. Rubén Rumbaut and Wayne A. Cornelius (San Diego: Center for U.S.-Mexican Studies, University of California, 1995), 71–76; Rubén Rumbaut, "The Crucible Within: Ethnic Identity, Self-Esteem and Segmented Assimilation Among Children of Immigrants," in *The New Second Generation*, ed. Alejandro Portes (New York: Russell Sage Foundation, 1996), 119–70; Min Zhou, "Segmented Assimilation: Issues, Controversies, and Recent Research on the New Second Generation," *International Migration Review* 31 (Winter 1997): 975–1008; Alejandro Portes, Patricia Fernández-Kelly, and William Haller, "Segmented Assimilation on the Ground: The New Second Generation in Early Adulthood," *Ethnic and Racial Studies* 28 (November 2005): 1000–1040.
120. See Min Zhou and Carl Bankston III, "Social Capital and the Adaptation of the Second Generation: The Case of Vietnamese Youth in New Orleans," in *The New Second Generation*, ed. Alejandro Portes (New York: Russell Sage Foundation), 197–220; Alejandro Portes, "Social Capital"; Ricardo Stanton-Salazar, "A Social Capital Framework for Understanding the Socialization of Racial Minority Children and Youths," *Harvard Educational Review* 67 (Spring 1997): 1–40; Ricardo Stanton-Salazar, *Manufacturing Hope and Despair: The School and Kin Support Networks of U.S.-Mexican Youth* (New York: Teachers College Press, 2001).
121. Margaret Gibson, *Accommodation Without Assimilation: Sikh Immigrants in an American High School* (Ithaca: Cornell University Press, 1988).

122. Herbert Gans, "Second Generation Decline: Scenarios for the Economic and Ethnic Futures of the Post-1965 American Immigrants," *Ethnic and Racial Studies* 15 (March 1992): 173–92.
123. Alejandro Portes, Patricia Fernández-Kelly, and William Haller, "Segmented Assimilation."
124. Richard Alba and Victor Nee, "Rethinking Assimilation Theory for a New Era of Immigration," *International Migration Review* 31 (Winter 1997): 826–74; Frank D. Bean and Gillian Stevens, *America's Newcomers and the Dynamics of Diversity* (New York: Russell Sage Foundation, 2003).
125. Alejandro Portes, Patricia Fernández-Kelly, and William Haller, "Segmented Assimilation."
126. Reynolds Farley and Richard Alba, "The New Second Generation in the United States," *International Migration Review* 36 (Autumn 2002): 669–701.
127. Rubén Rumbaut, "Turning Points in the Transition to Adulthood: Determinants of Educational Attainment, Incarceration, and Early Childbearing among Children of Immigrants," *Ethnic and Racial Studies* 28 (November 2005): 1041–86.
128. Marcelo M. Suárez-Orozco and Carola Suárez-Orozco, "The Cultural Patterning of Achievement Motivation: A Comparison of Mexican, Mexican Immigrant, Mexican American, and Non-Latino White American Students," in *California's Immigrant Children*, 160–90; Alejandro Portes and Rubén Rumbaut, *Legacies*.
129. Joel Perlmann, *Italians Then, Mexicans Now*, passim.
130. Sharon L. Sassler, "School Participation Among Immigrant Youths: The Case of Segmented Assimilation in the Early 20th Century," *Sociology of Education* 79 (January 2006): 1–24.
131. *Ibid.*, 15.
132. Richard Alba, Amy Lutz, and Elena Vesselinov, "How Enduring Were the Inequalities Among European Immigrant Groups in the United States?" *Demography* 38 (August 2001): 349–56.
133. *Ibid.*, 355.
134. George Borjas, "Long-Run Convergence of Ethnic Skill Differentials: The Children and Grandchildren of the Great Migration," *Industrial and Labor Relations Review* 47 (July 1994): 553–73.
135. Laurie Olsen, *Made in America: Immigrant Students in Our Public Schools* (New York: The New Press, 1997); Laurie Olsen, "Public Education, Immigrants, and Racialization: The Contemporary Americanization Project," in *E Pluribus Unum? Contemporary and Historical Perspectives on Immigrant Political Incorporation*, ed. Gary Gerstle and John Mollenkopf (New York: Russell Sage Foundation, 2001), 371–411.
136. Alejandro Portes and Rubén Rumbaut, *Legacies*.
137. Mary Waters, "Ethnic and Racial Identities of Second-Generation Black Immigrants in New York City," in *The New Second Generation*, 171–96; Alejandro Portes and Rubén Rumbaut, *Legacies*; Philip Kasinitz, Juan Battle, and Ines Miyares, "Fade to Black: The Children of West Indian Immigrants in Southern Florida," in *Ethnicities: Children of Immigrants in America*, ed. Rubén G. Rumbaut and Alejandro Portes (Berkeley: University of California Press, 2001), 267–300.
138. Samuel P. Huntington, *Who Are We? The Challenges to America's National Identity* (New York: Simon and Schuster, 2004).
139. Matthew Frye Jacobson, *Whiteness of a Different Color: European Immigrants and the Alchemy of Race* (Cambridge, MA: Harvard University Press, 1998); James R. Barrett and David Roediger, "Inbetween Peoples: Race, Nationality, and the 'New Immigrant' Working Class," *Journal of American Ethnic History* 16 (Spring 1997): 3–44;

David R. Roediger, *Working Toward Whiteness: How America's Immigrants Became White—The Strange Journey from Ellis Island to the Suburbs* (New York: Basic Books, 2005).

140. Gary Gerstle, *American Crucible: Race and Nation in the Twentieth Century* (Princeton: Princeton University Press, 2001).

141. Paula Fass, *Outside In*, passim.

142. Frank D. Bean and Gillian Stevens, *America's Newcomers*.

143. Richard Alba and Victor Nee, "Rethinking Assimilation Theory"; Richard Alba and Victor Nee, *Remaking the American Mainstream: Assimilation and Contemporary Immigration* (Cambridge, MA: Harvard University Press, 2003).

144. Joel Perlmann and Roger Waldinger, "Second Generation Decline?"

145. Herbert Gans, "Second Generation Decline."

146. William J. Wilson, *The Declining Significance of Race: Blacks and Changing American Institutions* (Chicago: University of Chicago Press, 1978); Orlando Patterson, *The Ordeal of Integration: Progress and Resentment in America's "Racial" Crisis* (Washington, DC: Civitas/Counterpoint, 1997).

147. Philip Kasinitz, John Mollenkopf, and Mary C. Waters, "Becoming American/Becoming New Yorkers: Immigrant Incorporation in a Majority Minority City," *International Migration Review* 36 (December 2002): 1020–36.

148. Patricia Fernandez-Kelly and Richard Schauffler, "Divided Fates: Immigrant Children and the New Assimilation," in *The New Second Generation*, 30–53; Mary Waters, "Ethnic and Racial Identities"; Alex Stepick et al., "Shifting Identities and Intergenerational Conflict: Growing up Haitian in Miami," in *Ethnicities*, 229–66.

149. Alejandro Portes, "Social Capital," 6; see especially James S. Coleman, "Social Capital in the Creation of Human Capital," *American Journal of Sociology* 94 (Supplement, 1988): S95–S120.

150. Alejandro Portes and Rubén Rumbaut, *Legacies*.

151. Pedro Portes, "Social and Psychological Factors in the Academic Achievement of Children of Immigrants: A Cultural History Puzzle," *American Educational Research Journal* 36 (Fall 1999): 489–507; Michael R. Olneck, "Immigrants and Education."

152. Alejandro Portes and Dag MacLeod, "Educational Progress of Children of Immigrants: The Roles of Class, Ethnicity, and School Context," *Sociology of Education* 69 (October 1996): 255–75; Min Zhou and Carl Bankston III, *Growing Up American*; Pedro Portes, "Social and Psychological Factors"; Alejandro Portes and Rubén Rumbaut, *Legacies*.

153. Ricardo Stanton-Salazar, "A Social Capital Framework"; Ricardo Stanton-Salazar, *Manufacturing Hope and Despair*; David Lopez and Ricardo Stanton-Salazar, "Mexican Americans: A Second-Generation at Risk," in *Ethnicities*, 57–90.

154. Angela Valenzuela, *Subtractive Schooling: U.S.-Mexican Youth and the Politics of Caring* (Albany, NY: SUNY Press, 1999).

155. Reed Ueda, "Second-Generation Civic America"; Claudia Goldin and Lawrence F. Katz, "Human Capital and Social Capital: The Rise of Secondary Schooling in America, 1910–1940," *Journal of Interdisciplinary History* 29 (Spring 1999): 683–723; John L. Rury, "Social Capital and Secondary Schooling: Interurban Differences in American Teenage Enrollment Rates," *American Journal of Education* 110 (August 2004): 293–320.

156. Michael R. Olneck and Marvin F. Lazerson, "Immigrant School Achievement."

157. Morris Berger, "The Settlement, the Immigrant, and the Public School" (PhD diss., Columbia University, 1956).

158. Arthur A. Goren, *New York Jews and the Quest for Community: The Kehillah Experiment, 1908–1922* (New York: Columbia University Press, 1970).
159. Raymond A. Mohl, "Schools, Politics, and Riots."
160. Reed Ueda, "Second-Generation Civic America."
161. *Ibid.*, 662.
162. Rudolph J. Vecoli, "Contadini in Chicago: A Critique of *The Uprooted*," *Journal of American History* 51 (December 1964), 412–17.

CHAPTER 6

THE HISTORIOGRAPHY OF EDUCATION FOR GIRLS AND WOMEN IN THE UNITED STATES

Margaret A. Nash[1]

The historiography of female education in the United States grows out of two major fields, each with its own trajectory and contribution: women's history, and the history of education. In women's history, work has developed from a movement for inclusion of female experience in historical narratives, to a sociocultural approach that interrogates the meaning of gender itself. In education, work also has been transformed in the last few decades. As early as 1960, the historian Bernard Bailyn wrote that he hoped to see fewer institutional histories or histories of formal schooling, as they tended to be written as though schooling existed apart from the culture in which it was embedded. He challenged historians to write about education as a process of cultural transmission, in which a particular institution might play a specific role.[2] Research on education, including the history of female education, has heeded this call. In this chapter, I provide a brief discussion of trends in women's history, with examples of how women's educational history has built on these trends. Most notably, recent historiography on female education includes discussions of class, race, and gender identity, and the roles of education in creating or altering those identities; transnational perspectives; and challenges both to periodization and to the presumed rationales for female education. I conclude with thoughts about promising avenues for future research.

TRENDS IN WOMEN'S HISTORY AND EDUCATIONAL HISTORY

The modern field of women's history emerged in the 1960s. As Nancy Cott outlined in a recent retrospective on the state of the field, the initial goal of women's history was to render women visible, given the near total absence of any inclusion of women or women's contributions until then.[3] This work was fostered by new social histories, which moved beyond political and military histories to include far more historical actors, conditions, and perspectives. Several historians created theoretical models for the growth of women's history. The most familiar is that of Gerda Lerner and Mary Kay Thompson Tetreault's model, which moves from male-defined history, to compensatory history (often referred to as the "add women and stir" approach, because it added a few notable women to the existing narrative without changing or challenging any of the basic assumptions), to contribution history (exploring women's contributions to society), to histories that analyze the framework of oppression in which women lived, and finally to histories with a female consciousness.[4]

The first major comprehensive work on the history of women's education since the revisionists, Barbara Miller Solomon's *In the Company of Educated Women: A History of Women and Higher Education in America*, fits the early goal of inclusion.[5] Solomon provided a broad synthesis of women's education, documented women's fight for access to institutions and for fair treatment once admitted to colleges and universities, looked at opportunities for non-elite and immigrant women, took academies and seminaries seriously as institutions providing higher education, and investigated the opposition to increased access to education for women. A plethora of studies built on Solomon's work, adding rich detail about women's experiences in particular institutions, as students, faculty, and administrators.

There were three main challenges to women's history in the 1980s and 1990s. First, postmodern theories of the "subject" questioned the authoritativeness of texts, and led historians to look at documents and narratives in new ways. Second, scholars challenged the implicit assumption of whiteness in the early work on women. Third, some scholars argued that the focus needed to be on "gender" rather than on "women's" history, contending that an essential piece of the work ought to be on understanding gender as a construct and as part of a discourse of power. This opened up the possibility for studies of masculinity, as well, instead of assuming an understanding of historical maleness.[6] All of these challenges have in common at least one key element, and that is a shift in thinking about the category "woman." Where earlier studies privileged femaleness as a category of analysis and emphasized women's similarities, more recent work emphasized the situatedness of concepts of gender. Where the earlier scholarship tended to erase difference and see women as a group, newer scholarship looked closely at differences of race, class, and ethnicity, and produced more nuanced histories. Finally, newer scholarship understands gender discourse as dialectical, insisting that we cannot understand femaleness in a particular time or place without also analyzing maleness.

Some of these trends have found their way into histories of female education. This is especially true regarding socioeconomic class. While earlier histories focused implicitly on white middle- or upper-class girls and women, later histories were explicit about who was being studied, and turned their focus to understudied groups, such as immigrant girls and children in orphanages.[7] Work on reformatories is a good example of changes in the historiography. The revisionist approach employed a Marxist analysis, arguing that reformatories were a way for the elite to control the working class. Other historians looked at race and ethnicity; Douglas Wertsch found that a disproportionate number of the inmates in Iowa's reform school for girls in the late nineteenth century were immigrants or African Americans. Feminist historical approaches, such as that of Barbara Brenzel and Ruth Alexander, saw agency on the part of working class parents, many of whom sent their children to state institutions. Mary Odem highlighted the complexities and contradictions among women reformers, as well as uncovering the voices of some of the girls deemed delinquent, concluding that however well intentioned reformers might have been, they ultimately further stigmatized working class girls.[8]

Other researchers have explored the relationship between education and labor. Susan Carter and Mark Prus argued that girls outnumbered boys in Progressive Era high schools because it was one of the only sources of job training for girls.[9] John Rury documented regional variation in the rise of sex-differentiated vocational, commercial, and home economics courses in high schools after 1900.[10] Such studies have illuminated some of the different purposes and types of education, as well as the different meanings of education in the lives of students from economically disadvantaged backgrounds.

Within higher education, scholars have opened the field beyond the study of elite colleges. In a review essay in 1997, Linda Eisenmann urged historians to rethink Solomon's "implicit hierarchy" that favored elite women's colleges (especially the Seven Sisters[11]), considering instead the influences of other agencies, such as foundations, the government, and accrediting groups.[12] Researchers are examining a wider range of institutions, including Lynn Gordon's work on state colleges and universities and Christine Ogren's work on "normal" (teacher training) schools.[13] Maresi Nerad used organizational theory to analyze the process by which female-dominated academic departments within a university got marginalized or eliminated at the University of California, Berkeley.[14] These and other studies, then, move beyond the issue of access and use the lens of gender to examine campus life, both for students and faculty.

The trend away from studying elite institutions has led to a growing body of literature on antebellum female academies and seminaries. First discussed primarily as precursors to the postbellum women's college movement, and then analyzed for the existence or lack of proto-feminist qualities, in the last decade or so scholars have examined these institutions on their own terms.[15] Separately and in tandem, Nancy Beadie and Kim Tolley have made enormous contributions in this field. Beadie traced the economic influences on the growth of the curricula in academies in New York, and found that state support did result in the promotion of female education to an extent. In the long-term, however, once academies had been folded into a tax-supported public school system, standardization meant the loss of educational innovation and alternative visions.[16] In other work, Beadie and Tolley make a compelling argument for reclaiming the importance of academies in the history of U.S. education, and situating female education within that context.[17]

These studies fill important gaps and add to our knowledge by broadening the types of institutions studied. "Women's education" can no longer be equated with a few elite institutions, as it now is well documented how many other sorts of institutions women attended, including public and private coeducational colleges, academies and seminaries, normal schools, and community colleges.[18]

Scholarship on the elite women's colleges, including the Seven Sisters, continues to interest some researchers. These works place higher education in the context of social history, and are not the narrow institutional histories that Bailyn decried. Patricia A. Palmieri provided a rich analysis of the first generation of women faculty at Wellesley College, where there was a rare commitment to an all-female faculty who set out to create an intellectual community.[19] She described the pitfalls and costs of such a world, and documented the loss that occurred when the administration began to hire male faculty. Helen Lefkowitz Horowitz analyzed student life and culture at the Seven Sisters, ingeniously using architecture as a lens through which to see changing views of young womanhood. She exposed the cultural meaning behind, for instance, the "cottage system" of student living compared to dormitory life, discussing the way that particular architectural styles reveal a perceived need for protection of female students.[20] Contributors to an anthology on Harvard and Radcliffe addressed the changing meanings and roles assigned to gender over several centuries, as played out within and outside of these two institutions.[21]

All of these changes in direction in the study of female education are parallel to the shifts in women's history toward studying differences among women, rather than studying females as an undifferentiated group. In addition to paying closer attention to differences of socioeconomic class, historians in the last several decades have added much to our knowledge of the educational experiences of different races, nationalities, and ethnic groups. Regarding the education of women of color, we know far more about African Americans than any other group.[22] Scholars have examined historically black women's colleges, such as Yolanda Watson's work on the early presidents of Spelman College; specific African American educators, such as Linda Perkins' work on Lucy Diggs Stowe, and Audrey McCluskey's on Mary McLeod Bethune; and African American women's struggle to enter particular professions, including Georgia Burnette's study of nursing.[23] Gunja SenGupta analyzed issues of race, class, and regional identity in the context of an African American benevolent association that established an orphanage and industrial school for the children of Southern migrant domestic workers in Progressive Era New York.[24] Excellent work has been done on the history of segregation of African Americans, especially in the reassessments that were inspired by the fiftieth anniversary of the *Brown v. Board of Education* decision. There is room for gendered analyses of the effects of segregation as well as of the end of *de jure* segregation. Valora Washington and Joanna Newman analyzed gender disparity in high school graduation rates since 1976 and concluded that the disparity reflected a decline for black men rather than an advancement for black women.[25]

Suellen Hoy examined the intersection of race and religion in her study of a Catholic girls' school in Chicago in the 1950s, while Carol Devens interrogated a different manifestation of those forces in her work on missionary day schools for Native American girls.[26] Devens found that assumptions about gender dynamics were crucial to these missionaries, as they shifted their educational focus to young girls as a means through which to influence male citizens. The juncture of race and class is analyzed in the experience of Native American girls and women at off-reservation boarding schools in Devon A. Mihesuah's history of the Cherokee Female Seminary.[27] Mihesuah argues that this seminary was a prime example of education as acculturation, and in particular, that the seminary perpetuated class divisions. The school was supported by wealthier Cherokees whose children were offered a liberal arts education, in contrast to the vocational education given to the majority of children who came from poorer Cherokee families.

Researchers have begun to look at the history of education for Latinas and for Asian American women. Critical work has been done on the history of bilingual education, and there is scope for examining this topic through the lens of gender. Studies of the activists who have promoted expanded educational opportunities for underrepresented groups of students have led to some gender analysis. Delores Delgado focused on Chicana student activism within the Los Angeles school "blowouts" of 1968.[28] As Eileen Tamura noted, scholarly work on Asian American educational history is a wide-open field.[29] When we further narrow our focus to gender and Asian American educational history, we see that the need for future work is great indeed. Much more can, and likely will, be done in the future for these and other groups of women.

In addition to renewed focus on race, class, and ethnicity, Kathryn Kish Sklar identifies another trend within women's history. This trend involves historical and gendered analysis of social structures and of American politics. Sklar notes that most studies by historians of women in the 1970s were of private life, but by the 1990s most such studies were of public life.[30] Elisabeth Hansot and David Tyack exemplify this approach in the history of education. In their history of coeducation, they ask how the institutional history of public schools looks when viewed through the lens of gender.[31] Their nuanced analysis demonstrates the many "ironies and obscure trans-formations" that are involved in the study of gender and education, providing a rich and complex base for future exploration.[32]

Histories of curriculum also have included gender analyses. This is especially noticeable in recent work on the history of science edu-cation. Kim Tolley showed the changes in science education for girls and women in all types of institutions, including academies and high schools, from the early national period to the present. In a major challenge to the belief held by many today that science is inherently a "boys' subject," she demonstrated that in the early nineteenth cen-tury, more girls than boys studied science, and a greater percentage of girls' schools offered physics, astronomy, and chemistry than did boys' schools.[33] Sevan Terzian documented the mixed messages regarding science received by high school girls in the post-World War II era. Based on his analysis of a high school science magazine, Terzian concludes that the editors encouraged teachers to promote careers in science for both girls and boys, but that this encourage-ment was mitigated by a dominant image in the magazine of scien-tists as men, and of women only in supportive roles.[34]

Questions of the meaning of gender itself have captured the attention of many historians in recent years. Joan Jacobs Brumberg and Miriam Formanek-Brunell have applied cultural history and social theory to the study of girlhood, explicating the ways that com-mercialism helped create specific constructs of gender identity for girls.[35] Anthony Rotundo, Michael Kimmel, Gail Bederman, and others have looked at the historical construction of maleness.[36] Such studies have found resonance in several recent studies of education. Jane Hunter argues that high schools of the late nineteenth century helped create a new type of adolescent: the young lady, instead of the girl—ironically, an unintended consequence of the relative equality experienced by girls in these schools helped reshape yet again the concept of gender as these girls went on to create new roles for women.[37] The growth of gender history has led to interesting work on understanding cultural meanings of masculinity, as well. Leon Jackson exemplifies this approach in the field of the history of higher education with his work on the social meanings of brotherhood and manhood at Harvard in the eighteenth century.[38]

If early work largely was tied to the issue of access, later work has been about what that education *meant* to those who did or did not gain access to it. How did education shape individuals' or groups' identities, including those based on gender? The movement toward gender history is reflected in works that ask how education was a marker of gender, race, or class identity. In this way, the field has expanded not only to include a range of types of formal and informal educational settings, it also has expanded from the issue of access to broader issues of cultural understanding and meaning-making. Nash's study of antebellum seminaries and academies concludes that—because advanced education was linked to the formation and consolidation of the newly forming white middle classes—educational distinctions between white men and women were not as pronounced as were class- and race-based distinctions.[39]

Within the field of women's history, many exciting new works have taken a transnational approach. Historians have identified trans-Atlantic influences in women's movements; analyzed the diffusion of U.S. feminism abroad and the influence of immigrant women on U.S. activism; and looked at the impact of international politics and global economies on women's lives and on ideologies of gender.[40] A similar movement occurred within the realm of gender and educational histories. Noriko Ishii analyzed the impact of American Progressivism in the work of two female missionaries to Japan who concentrated on the education of Japanese women.[41] Another example of a study of the influence of a U.S. female educator abroad is Angelo Repousis' study of Emma Willard and the Troy Society for the Advancement of Female Education in Greece. Repousis argues that Willard used the language of republicanism to promote a wide sphere of activism for women.[42] Eiichiro Azuma examined the education of a group of students born in the United States to Japanese parents, and educated in Japan in the decade prior to World War II.

The transnational education of thousands of such students was supported by the Japanese government, which hoped to educate American students who would be cultural bridges between the two countries. This education was clearly gendered, as young men took classes in history and politics, while young women took classes in how to conduct a tea ceremony, Japanese cooking, and etiquette.[43] Historians of higher education long have been aware of the European—and especially German—influences on university reform in the United States, and recently work has been done to begin to understand women's experiences. Barred from doctoral degree granting programs in the United States, women turned to Europe, especially Switzerland and Germany. Sandra Singer provides extraordinary archival evidence of the lives and career paths of over 1,300 such women, and Helen Lefkowitz Horowitz provides rich detail of one woman's experience in her masterful biography of M. Carey Thomas.[44] James Albisetti investigated European views of female education in the United States, finding that supporters of higher education in France, England, and Germany saw U.S. women's colleges as second rate, and therefore promoted coeducation in Europe; at the secondary level, however, some women teachers in Britain feared following the U.S. pattern of coeducation.[45]

Over twenty years ago, Ellen Condliffe Lagemann urged that future research go beyond examining women as students and teachers, asking that instead research be combined with methodological rethinking in order to produce new categories for synthesis, and urging that we look to education that was other than school-centered. We might ask altogether new sets of questions when we do this, such as, have school-centered histories led to misperceptions of the role of education in women's lives?[46] Rather than study women as educators or study institutions, Lagemann studied the role of education in the lives of a group of Progressive Era reformers and activists.[47] Eisenmann has echoed the call to study education in non-formal settings.[48] Similarly, Sally Schwager suggests we look at "counter-institutions" that women developed when denied entry into colleges and/or public arenas in which to act.[49]

Some of the non-formal settings that have garnered recent scholarly attention are reading circles, libraries, and women's magazines.[50] In this work, historians have found new information regarding reading patterns, and the meaning that education held for some women. For instance, Ronald and Mary Zboray discovered that, instead of women reading novels while men read newspapers, antebellum women and men read much the same literature.[51] Andrea Walton and others are examining the role of philanthropy in the development of women's education.[52] This work has enriched our understanding of the forms of education available to women and has broadened the scope of historical inquiry.

Questions of periodization regularly resurface among historians, and women's historians are no exception. New information and new questions result in reframing our understanding of particular time periods. Earlier generations of women's historians described the activism of the suffrage movement as the "First Wave" of feminism, and asserted that after the vote was won in 1920, women's activism decreased significantly until the women's liberation movement of the 1960s. Historians labeled the period of inactivity between the 1920s and the 1960s as the "doldrums," and the 1960s movement as the beginning of the "Second Wave." Recent scholarship has challenged this trajectory. Scholars have documented the work of feminists in labor unions, socialist causes, and in policymaking (for instance in the federal Women's Bureau) throughout the period of the "doldrums."[53] Feminist activism did not disappear in the interwar and prewar period—it merely took different forms.

Historians of women's education have joined the effort to rethink the image of women's activism. Until recently, women's education was thought to have slipped into its own period of doldrums, especially immediately following World War II. Because the G.I. Bill brought enormous changes to college campuses beginning in the late 1940s, and because histories are replete with the tale of how "Rosie the Riveter" returned home from the factory after the war, the emphasis in scholarship on this period has been on male students. On campuses across the country, the percentage of students who were female dropped to about one-third. Yet, as Linda Eisenmann shows, the actual numbers of female students continued to increase, and campuses were looking at how to accommodate their needs. How is it, Eisenmann asks, that in a period of alleged listlessness, a college president could have referred to "an explosion of interest, attention, and research about women in the 1950s"? Her work addresses this question, and offers a clear challenge to old assumptions.[54]

THINKING ABOUT THE PAST AND THE FUTURE

One trend that cuts across the fields of women's history and history of women's education is the preponderance of studies from the Progressive Era to the present. Far less has been done on colonial history, whether it is about higher learning, or about gender. Gerda Lerner analyzed over seven hundred items, including articles, books, and dissertations, published from 1998 through 2000 on various aspects of women's history, and found that half dealt with the twentieth century, and only 6 percent dealt with the colonial era.[55] When I searched the "America: History and Life" database for articles or books on women's education published between 2000 and the first half of 2006, I found 85 pieces, 55 (nearly 65 percent) of which were studies from the Progressive Era to the present; only 2 were on colonial history, while 25 were nineteenth century, and 3 were broad syntheses crossing time periods. In the past few years, more scholars have turned their attention to the early nineteenth century—for both women's and men's educational histories—and we can hope for more work on these time periods.

We also will continue to revise the historiography itself. One prominent current narrative is that histories of women's education began in the 1960s and 1970s with the flowering of the modern women's movement and the beginning of women's history as a distinct field. According to that story, the first histories were about women's lack of access to education, and women's oppression. The next step in the conceptual process was "woman-centered" history, in which the very things that oppressed women, such as their social segregation, resulted in empowerment; "separate spheres" created powerful bonding experiences, and a sense of identity that became important for subsequent political activism. However, we may need to rethink both the beginning of this scholarship and the periods in which this work developed. For example, in response to Title IX, scholarship of the 1970s focused on women's access to public institutions. But Maxine Schwartz Seller has pointed out that this was not the first time this kind of scholarship was done: Thomas Woody's landmark book, *A History of Women's Education in the United States*, published in 1929, focused on the same issue. What was the impetus for Woody's work, and how did it differ from later studies that discussed women's exclusion? Are there other examples of this type of scholarship?

In a related vein, the theories explaining the rationales for women's limited access to education, such as "republican motherhood" and "separate spheres," also came of age in the 1970s as historians focused on visibility and inclusion. In the broader field of women's history, these concepts have been refined and similar work must take place in the history of women's education. When the focus of inquiry was that of access, "republican motherhood" and "separate spheres" were reasonable answers to the question of why women's access to education was limited. Now that the nature of these ideologies has been challenged, their use as explanatory theories also needs to change.[56] They can be used as springboards for new questions. How was "separate spheres" used in a particular context? What did it mean to specific women and men in particular times and places? How did this concept hold different significance depending on class, race, region, or religion?

The increased focus on subjectivity in social history raises fruitful areas for study in women's education as well. Much work has been done of late on the meaning of citizenship and civic identity. Less has been done on the relationships among the rights of citizenship, education, and concepts of gender and race. Christine Ogren demonstrated that female normal school students in the late nineteenth century did not associate education with women's rights.[57] Did women at other educational institutions look at this issue differently? To what extent did social class have an impact on whether women saw education as separate from, or closely aligned with, social and political rights? When and why did other marginalized groups link their rights of citizenship to their access to education?

As a field, the history of women's education has burgeoned in the last thirty years. While historians continue to explore new arenas within the field, they also are creating new theoretical frameworks for understanding those histories. Influenced by social, cultural, intellectual, political, and economic history, the field has grown richer. As new studies are added, we can hope the field will grow even more nuanced and more theoretically sophisticated.

NOTES

1. Thank you to Jennifer de Forest, Lisa Levenstein, Bill Reese, John Rury, John Thelin, and Kim Tolley for their helpful comments and suggestions.
2. Bernard Bailyn, *Education in the Forming of American Society* (Chapel Hill: University of North Carolina Press, 1960).
3. Nancy Cott et al., "Considering the State of U.S. Women's History," *Journal of Women's History* 15 (Spring 2003): 145.
4. Gerda Lerner, *The Majority Finds Its Past: Placing Women in History* (New York: Oxford University Press, 1981); Mary Kay Thompson Tetreault, "Integrating Women's History: The Case of United States History High School Textbooks," *The History Teacher* 19 (February 1986): 211.
5. Barbara Miller Solomon, *In the Company of Educated Women: A History of Women and Higher Education in America* (New Haven: Yale University Press, 1985).
6. Cott et al., "Considering the State of U.S. Women's History, " 147, 49.
7. David R. Contosta, *Philadelphia's Progressive Orphanage: The Carson Valley School* (University Park: Pennsylvania State University Press, 1997); Sheldon Hanft, "Mordecai's Female Academy," *American Jewish History* 79 (Spring1989): 72–93; Melissa Klapper, "'A Long and Broad Education': Jewish Girls and the Problem of Education in America, 1860–1920," *Journal of American Ethnic History* 22 (Fall 2002): 3–31; Nancy B. Sinkoff, "Educating for 'Proper' Jewish Womanhood: A Case Study in Domesticity and Vocational Training, 1897–1926," *American Jewish History* 77 (December 1988): 572; Robert S. Wolff, "Industrious Education and the Legacy of Samuel Ready, 1887–1920," *Maryland Historical Magazine* 95 (Fall 2000): 308–29.
8. Ruth Alexander, *The Girl Problem: Female Sexual Delinquency in New York, 1900–1930* (Ithaca: Cornell University Press, 1995); Barbara Brenzel, *Daughters of the State: A Social Portrait of the First Reform School for Girls in North America, 1856–1905* (Cambridge, MA: MIT Press, 1983); Georgina Hickey, "Rescuing the Working Girl: Agency and Conflict in the Michigan Reform School for Girls, 1879–1893," *Michigan Historical Review* 20 (Spring 1994): 1–28; Mary Odem, *Delinquent Daughters: Protecting and Policing Adolescent Female Sexuality in the United States, 1885–1920* (Chapel Hill: University of North Carolina Press, 1995); Douglas Wertsch, "Iowa's Daughters: The First Thirty Years of the Girls Reform School of Iowa, 1869–1899," *Annals of Iowa* 49 (Spring 1987): 77–100.
9. Susan B. Carter and Mark Prus, "The Labor Market and the American High School Girl, 1890–1928," *Journal of Economic History* 42 (March1982): 172.
10. John L. Rury, *Education and Women's Work: Female Schooling and the Division of Labor in Urban America, 1870–1930* (Albany: State University of New York Press, 1991).
11. The "Seven Sisters" are Mount Holyoke, Vassar, Smith, Wellesley, Radcliffe, Bryn Mawr, and Barnard colleges.
12. Linda Eisenmann, "Reconsidering a Classic: Assessing the History of Women's Education a Dozen Years after Barbara Solomon," *Harvard Educational Review* 67, no. 4 (1997): 694.
13. Lynn Gordon, *Gender and Higher Education in the Progressive Era* (New Haven: Yale University Press, 1990); Jurgen Herbst, *And Sadly Teach: Teacher Education and*

Professionalization in American Culture (Madison: University of Wisconsin Press, 1989); Christine A. Ogren, *The American State Normal School: "An Instrument of Great Good"* (New York: Palgrave Macmillan, 2005).

14. Maresi Nerad, *The Academic Kitchen: A Social History of Gender Stratification at the University of California, Berkeley* (Albany: State University of New York Press, 1999).

15. David F. Allmendinger, Jr., "Mount Holyoke Students Encounter the Need for Life Planning, 1837–1850," *History of Education Quarterly* 19 (Spring 1979): 27; Kathryn M. Kerns, "Farmers' Daughters: The Education of Women at Alfred Academy and University before the Civil War," *History of Higher Education Annual* 6 (1986); Keith Melder, "Masks of Oppression: The Female Seminary Movement in the United States," in *History of Women in the United States: Historical Articles on Women's Lives and Activities,* ed. Nancy F. Cott (Munich, New Providence, London, and Paris: K. G. Saur, 1993); Anne Firor Scott, "The Ever-Widening Circle: The Diffusion of Feminist Values from the Troy Female Seminary, 1822–1872," *History of Education Quarterly* 19 (Spring 1979): 3–25.

16. Nancy Beadie, "Emma Willard's Idea Put to the Test: The Consequences of State Support of Female Education in New York, 1819–67," *History of Education Quarterly* 33 (Winter 1993): 543.

17. Kim Tolley, "Mapping the Landscape of Higher Schooling, 1727–1850," in *Chartered Schools: Two Hundred Years of Independent Academies in the United States, 1727–1925,* ed. Nancy Beadie and Kim Tolley (New York and London: RoutledgeFalmer, 2002); Kim Tolley and Nancy Beadie, "A School for Every Purpose: An Introduction to the History of Academies in the United States," in *Chartered Schools: Two Hundred Years of Independent Academies in the United States, 1727–1925,* ed. Nancy Beadie and Kim Tolley (New York and London: RoutledgeFalmer, 2002).

18. John H. Frye, "Women in the Two-Year College, 1900 to 1970," *New Directions for Community Colleges* 89 (Spring 1995): 5–14.

19. Patricia Ann Palmieri, *In Adamless Eden: The Community of Women Faculty at Wellesley* (New Haven: Yale University Press, 1995).

20. Helen Lefkowitz Horowitz, *Alma Mater: Design and Experience in the Women's Colleges from Their Nineteenth-Century Beginnings to the 1930s* (Amherst: University of Massachusetts Press, 1984).

21. Laurel Thatcher Ulrich, ed., *Yards and Gates: Gender in Harvard and Radcliffe History* (New York: Palgrave Macmillan, 2004).

22. Linda Eisenmann, "Creating a Framework for Interpreting U.S. Women's Educational History: Lessons from Historical Lexicography," *History of Education* 30 (Winter 2001): 455–56.

23. Georgia Burnette, "Looking Back: Black Nurses Struggle for Admission to Professional Schools," *Afro-Americans in New York Life and History* 28 (Summer 2004): 85; Lauri Johnson, "A Generation of Women Activists: African American Female Educators in Harlem, 1930–1950," *Journal of African American History* 89 (Summer 2004): 223; Audrey Thomas McCluskey, "Multiple Consciousness in the Leadership of Mary Mcleod Bethune," *NWSA Journal* 6 (Spring 1994): 69; Linda Perkins, "Lucy Diggs Slowe: Champion of the Self-Determination of African American Women in Higher Education," *Journal of Negro History* 81 (Winter 1996): 89; Linda Perkins, "The Role of Education in the Development of Black Feminist Thought, 1860–1920," *History of Education* 22, no. 3 (1993): 265; Stephanie J. Shaw, "'We Are Not Educating Individuals but Manufacturing Levers': Creating a Black Female Professional Class During the Jim Crow Era," *Magazine of History* 18

(April 2004): 17–21; Yolanda Watson, *Daring to Educate: The Legacy of the Early Spelman College Presidents* (Sterling: Stylus, 2005); Heather Andrea Williams, *Self-Taught: African American Education in Slavery and Freedom* (Chapel Hill: University of North Carolina Press, 2005).

24. Gunja SenGupta, "Elites, Subalterns, and American Identities: A Case Study of African-American Benevolence," *American Historical Review* 109 (October 2004): 1104.

25. Valora Washington and Joanna Newman, "Setting Our Own Agenda: Exploring the Meaning of Gender Disparities among Blacks in Higher Education," *Journal of Negro Education* 60 (Spring 1991): 19–35.

26. Carol Devens, "'If We Get the Girls, We Get the Race': Missionary Education of Native American Girls," *Journal of World History* 3 (June 1992): 219; Suellen Hoy, "No Color Line at Loretto Academy: Catholic Sisters and African Americans on Chicago's South Side," *Journal of Women's History* 14 (Spring 2002): 8–33.

27. Amanda J. Cobb, *Listening to Our Grandmothers' Stories: The Bloomfield Academy for Chickasaw Females, 1852–1949* (Lincoln: University of Nebraska Press, 2000); Devon A. Mihesuah, *Cultivating the Rosebuds: The Education of Women at the Cherokee Female Seminary, 1851–1909* (Urbana: University of Illinois Press, 1993, 1998).

28. Dolores Bernal Delgado, "Grassroots Leadership Reconceptualized: Chicana Oral Histories and the 1968 East Los Angeles School Blowouts," *Frontiers* 19, no. 2 (1998): 113.

29. Eileen H. Tamura, "Asian Americans in the History of Education: An Historiographical Essay," *History of Education Quarterly* 41 (Spring 2001): 58–71; Eileen H. Tamura, "Introduction: Asian Americans and Educational History," *History of Education Quarterly* 43 (Spring 2003): 3.

30. Cott et al., "Considering the State of U.S. Women's History," 147.

31. Elisabeth Hansot and David Tyack, "Gender in American Public Schools: Thinking Institutionally," *Signs: Journal of Women in Culture and Society* 13 (Summer 1988): 741–60; David Tyack and Elisabeth Hansot, *Learning Together: A History of Coeducation in American Public Schools* (New Haven: Yale University Press, 1990).

32. Tyack and Hansot, *Learning Together*, 12.

33. Kim Tolley, *The Science Education of American Girls: A Historical Perspective* (New York and London: RoutledgeFalmer, 2003).

34. Sevan G. Terzian, "*Science World*, High School Girls, and the Prospect of Scientific Careers, 1957–1963," *History of Education Quarterly* 46 (Spring 2006): 73.

35. Joan Jacobs Brumberg, *The Body Project: An Intimate History of American Girls* (New York: Vintage, 1998); Miriam Formanek-Brunell, *Made to Play House: Dolls and the Commercialization of American Girlhood, 1830–1930* (New Haven: Yale University Press, 1993).

36. Gail Bederman, *Manliness & Civilization: A Cultural History of Gender and Race in the United States, 1880–1917* (Chicago: University of Chicago Press, 1995); Michael S. Kimmel, *Manhood in America: A Cultural History* (New York: Free Press, 1996); E. Anthony Rotundo, *American Manhood: Transformations in Masculinity from the Revolution to the Modern Era* (New York: Basic Books, 1993).

37. Jane H. Hunter, *How Young Ladies Became Girls: The Victorian Origins of American Girlhood* (New Haven: Yale University Press, 2002).

38. Leon Jackson, "The Rights of Man and the Rites of Youth: Fraternity and Riot at Eighteenth-Century Harvard," in *The American College in the Nineteenth Century*, ed. Roger L. Geiger (Nashville: Vanderbilt University Press, 2000).

39. Margaret A. Nash, *Women's Education in the United States, 1780–1840* (New York and London: Palgrave, 2005).

40. Mari Yoshihara, "Tackling the Contested Categories: Culture, Race, and Nation in American Women's History," *Journal of Women's History* 15 (Spring 2003): 164.

41. Noriko Ishii, "Crossing Boundaries of Womanhood: Professionalization and American Women Missionaries' Quest for Higher Education in Meiji Japan," *Journal of American and Canadian Studies* 19 (2001): 85–122.

42. Angelo Repousis, "The Trojan Women: Emma Hart Willard and the Troy Society for the Advancement of Female Education in Greece," *Journal of the Early Republic* 24 (Fall 2004): 445.

43. Eiichiro Azuma, 'The Pacific Era Has Arrived': Transnational Education Among Japanese Americans, 1932–1941," *History of Education Quarterly* 43 (Spring 2003): 39–71.

44. Helen Lefkowitz Horowitz, *The Power and Passion of M. Carey Thomas* (New York: Alfred A. Knopf, 1994); Sandra L. Singer, *Adventures Abroad: North American Women at German-Speaking Universities, 1868–1915* (New York: Praeger, 2003).

45. James C. Albisetti, "American Women's Colleges through European Eyes, 1865–1914," *History of Education Quarterly* 32 (Winter 1992): 439; James C. Albisetti, "Un-Learned Lessons from the New World? English Views of American Coeducation and Women's Colleges, c. 1865–1910," *History of Education* 29, no. 5 (2000): 473.

46. Ellen Condliffe Lagemann, "Looking at Gender: Women's History," in *Historical Inquiry in Education: A Research Agenda*, ed. John Hardin Best (Washington: American Educational Research Association, 1983), 252–60.

47. Ellen Condliffe Lagemann, *A Generation of Women: Education in the Lives of Progressive Reformers* (Cambridge, MA: Harvard University Press, 1979).

48. Eisenmann, "Creating a Framework."; Eisenmann, "Reconsidering a Classic."

49. Eisenmann, "Reconsidering a Classic," 698.

50. Mary Kelley, "'A More Glorious Revolution': Women's Antebellum Reading Circles and the Pursuit of Public Influence," *New England Quarterly* 76 (June 2003): 163; Laura McCall, "'The Reign of Brute Force Is Now Over': A Content Analysis of *Godey's Lady's Book*, 1830–1860," *Journal of the Early Republic* 9 (Summer 1989): 217.

51. Ronald J. Zboray and Mary Saracino Zboray, "Home Libraries and the Institutionalization of Everyday Practices among Antebellum New Englanders," *American Studies* 42 (Fall 2001): 63–86.

52. Andrea Walton, ed., *Women and Philanthropy in Education* (Bloomington: Indiana University Press, 2005).

53. Susan M. Hartmann, *The Other Feminists: Activists in the Liberal Establishment* (New Haven: Yale University Press, 1998); Dawn Keetley and John Pettegrew, eds., *Public Women, Public Words: A Documentary History of American Feminism* (Lanham: Rowman & Littlefield, 2002); Joanne Meyerowitz, ed., *Not June Cleaver: Women and Gender in Postwar America, 1945–1960* (Philadelphia: Temple University Press, 1994); Shaw, "'We Are Not Educating Individuals but Manufacturing Levers': Creating a Black Female Professional Class During the Jim Crow Era."; Stephanie J. Shaw, *What a Woman Ought to Be and to Do: Black Professional Women Workers During the Jim Crow Era, Women in Culture and Society* (Chicago: University of Chicago Press, 1996).

54. Linda Eisenmann, *Higher Education for Women in Postwar America, 1945–1965* (Baltimore: The Johns Hopkins University Press, 2006), 1–4.

55. Cott et al., "Considering the State of U.S. Women's History," 146–47.

56. Margaret A. Nash, "Rethinking Republican Motherhood: Benjamin Rush and the Young Ladies' Academy of Philadelphia," *Journal of the Early Republic* 17 (Summer 1997): 171–91.
57. Ogren, *The American State Normal School*, passim.

CHAPTER 7

CHILDREN IN AMERICAN HISTORY

N. Ray Hiner

The history of children is a relatively new field compared to the history of education, gaining its original impetus from the work of social historians and psychohistorians in the 1960s. When I first encountered the history of children in the early 1970s, it was a nascent field, full of promise, but undeveloped and scattered across a variety of disciplines and specialties. Today, more than thirty years later, the field has emerged as a robust, multidisciplinary enterprise with its own professional organization and a new scholarly journal. In 2001, after initial support from the Benton Foundation, a group of scholars established the Society for the History of Children and Youth. In 2007, the Society will hold its Fourth Biennial Conference in Sweden. In June 2007, the first issue of the new *Journal of the History of Children and Youth* is scheduled to appear.[1]

Another illustration of the vitality of the field is the recent appearance of several synthetic works, anthologies, and collections. Among the most recent surveys are those by Steven Mintz (*Huck's Raft: A History of American Childhood*, 2004), Joseph Illick (*American Childhoods*, 2002), Elliott West (*Growing Up in Twentieth-Century America*, 1996), and Harvey Graff (*Conflicting Paths: Growing Up in America*, 1995). In addition, Anya Jabour's useful anthology, *Major Problems in the History of American Families and Children*

(2005), the extensive collection of documents and essays, *Childhood in America* (2000) edited by Paula S. Fass and Mary Ann Mason, and *Small Worlds: Children & Adolescents in America, 1850–1950* (1992), a collection of original essays edited by Elliott West and Paula Petrik, represent important contributions to the field.[2] All of these developments reflect the remarkable expansion in the scope and the striking improvement in the quality of the scholarship on the history of children and youth that has occurred in the last two decades. Indeed, an impressive number of recent scholars have been recognized for their work.[3] There is little reason to doubt that this field will continue to develop and expand its influence.[4]

In retrospect, I realize that I was drawn to the history of children in large measure because, as a young historian of education, I sensed that to study children, to really study children, was fundamental to understanding the human condition, and that to study children's history was essential to understanding human history itself. To state the obvious, children are basic to human existence, and to experience childhood is the lot of every human being. Or as Mark Twain put it, "We haven't all had the good fortune to be ladies, we haven't all been generals, or poets, or statesmen, but when the toast works down to the babies, we all stand on common ground."[5]

Perhaps it is unnecessary to argue here that this "common ground"—the universal experience of childhood—has important implications for the history of education. Children are obviously central to the process of education, however defined. Perhaps more than most historians, we historians of education have appreciated the unique perspective that can be gained by the study of children, past and present. Perhaps—but there is ample evidence that even if we have generally given more attention to children than other historians, we have not fully utilized the considerable historical leverage that the study of children in the past offers to historians of education. Until recently, we have been so intensely occupied with the study of educational ideas and institutions, teachers and administrators, teaching methods and curricula, class and race, and gender and ethnicity that children did not usually come to center stage except to illustrate these important topics.[6] James Axtell observed more than two decades ago that historians of education too often failed to capture "some sense of what the educational process was like at the level at which it actually occurred." Axtell urged historians "to portray a waist-high view of education, one that enables us to see the educational process, if not actually through children's eyes, at least from their position in a Lilliputian universe."[7] Barbara Finkelstein used a different metaphor to make a similar point when she concluded that historians of education had too often "relegated learners and learning to the back seat of the historical bus." She encouraged her fellow historians of education "to incorporate an awareness of children into our understanding of modern educational history."[8] Since professors Axtell and Finkelstein offered their recommendations in the 1970s, children have indeed received more attention from historians of education, but it is clear that much work remains to be done.[9] Thus, what follows is a modest effort to review *some* of the major themes in the history of children in what is now the United States. The field is too complex and dynamic to be summarized adequately in this brief essay, but I hope it may be helpful to historians of education as they seek incorporate children more fully into the history of education.

CHILDREN IN EARLY AMERICA

Historians of children in early America (before 1800) have concentrated primarily on two areas: (1) basic demographic patterns such as birth and death rates, size of families, and age of marriage, and (2) adult attitudes toward children and related child-rearing practices. Children of native peoples were obviously central to the remarkable process by which humans populated the Americas. We have virtually no direct evidence of birth and mortality rates for these peoples, but demographers have concluded that there was a generally steady and at times explosive growth in human population throughout most of the long period before European contact. Henry Dobyns asserts that before European contact Native American peoples lived in an environment generally free of diseases that were endemic in Europe. Thus, children and adults "suffered no smallpox, no measles, no chickenpox, no influenza, no typhus, no typhoid, or paratyphoid fever, no diphtheria, no cholera, no bubonic plague, no scarlet fever."[10] However, Native Americans, like all pre-modern peoples, were susceptible to climatic and ecological changes that could bring hardship, starvation, and severe dislocations, sometimes resulting in the disappearance or relocation of entire communities.[11] Children were obviously caught up in these traumatic events along with their families and communities. Still, Native American children generally thrived before the European contact.

When the Europeans arrived, they brought death to Native American children, and with death came despair for the living. Spiritual disorientation, cultural disintegration and alienation, and physical displacement were too often the fate of those children and adults who managed to survive the consequences of contact with the Europeans. The basic facts of this demographic disaster that befell native peoples after European contact are well known, so I will not repeat them here.[12] Yet, to grasp the full meaning of this tragedy, it is important to keep in mind that this traumatic process was experienced by children as well as adults, by children who were not as well equipped developmentally to cope with these horrors as were adults. To be sure, tribal communities, even those under tremendous stress, provided critical emotional support to the children who survived, but they carried emotional scars with them throughout their lives as they continued to try to adjust to the increasing threat posed by the Europeans to their cultures and often to their lives. Throughout North America, in colonial New Mexico, in New England, as well as the Chesapeake and New France, Native American children were central to their cultures' survival. Gradually, the children who survived the consequences of the biological exchange and the cultural conflict that followed changed the complexion and character of their societies.[13]

Children of Europeans had their own demographic and cultural challenges to overcome, although some fared much better than others. For example, Anglo American children born in seventeenth-century New England had a remarkably good chance of surviving their first year of life by seventeenth-century European standards, although infant mortality rates of one in ten live births are high by modern standards. Moreover, most of these New England children could expect to come of age without the loss of a parent. Apparently, as many as 60–80 percent of seventeenth-century New England marriages remained unbroken by parental death until the normal period of childbearing was completed. When "premature" parental death did occur, the surviving parent usually remarried within a short time.[14]

Conditions for children in the seventeenth-century Chesapeake contrasted sharply with those in New England. According to Ross Beales, their situation was almost the reverse of that of the children in New England. In the Chesapeake, "life expectancy was low; completed families were unusual; parents rarely lived to see their grandchildren; children frequently lost one parent; and orphans were common."[15] In seventeenth-century Maryland, for example, perhaps one third of all male children died before they reached the age of twenty, and parental death was astonishingly high.[16] The consequences for the children and adults who survived these conditions were profound. Darrett and Anita Rutman found one household in seventeenth-century Virginia created by a chain of marriages, parental deaths, and remarriages that had produced at least twenty-five children. "A visit to this household in 1680 would have found the presence of children (ranging from infancy to the early twenties) from four of the marriages, some of whom did not have any parents in common!"[17] The Rutmans concluded "that we are dealing with two entirely different types of childhood along the seventeenth-century Anglo-American coast—in New England one lived more in a parental situation; in the Chesapeake area one lived more in a kinship situation."[18] Fortunately, by the middle of the eighteenth century, the basic conditions of life for Chesapeake children improved dramatically. Life expectancy increased, infant mortality dropped (although still high by modern standards), and a more stable family environment developed, with larger households, fewer broken and reconstituted families, and, happily, fewer orphans.[19]

Ironically, while conditions for Chesapeake children were improving in the eighteenth century, they were worsening for children in New England. A growing population increasingly concentrated in small towns, and cities such as Boston, made children highly susceptible to contagious diseases. Thousands of children fell victim to epidemics of smallpox, diphtheria, scarlet fever, and yellow fever that ravished New England during this period. John Duffy reported that in a diphtheria epidemic in the 1730s, more than one half of all the children in one community under the age of fifteen years died.[20] Still, mortality rates in eighteenth-century New England did not regress to seventeenth-century Chesapeake levels or even to those of contemporary Europe. Research on mortality trends in the middle colonies is less complete, but generally indicates that in the eighteenth-century mortality rates for this region fell somewhere between those for New England and the Chesapeake, except for Philadelphia and New York, which had rates similar to those of Boston.[21]

African American children experienced higher mortality rates than Anglo American children, although they generally fared better than Native American children.[22] In general, demographic trends for the Chesapeake slave population roughly paralleled that for whites in that mortality rates and sex ratios were highest during the seventeenth century, and improved dramatically during the eighteenth century. However, conditions were always worse for slave children than for white children, and improvement in mortality occurred later among the slave population than among whites.[23] Paradoxically, high mortality among slaves, especially in the seventeenth century—in concert with the refusal of colonial governments to grant legal recognition to slave marriages—encouraged the emergence of strong kin consciousness among slaves that in some ways resembled that among Chesapeake whites. For slaves, kin relationships provided important emotional support and physical assistance in time of great need. However, we should remember that children born into slavery were nothing more than property in the eyes of the law, and therefore, always vulnerable to mistreatment and abuse. Even so, the presence of supportive kin and an emerging African American community helped create an environment in which slave children could obtain the essential nurture and respect they needed as human beings. Still, this was only a possibility and not necessarily available to all or even most slave children most of the time. Children are highly adaptable, but slavery tested the limits of their adaptability.[24]

These basic facts of life and death in early America provide an essential framework for understanding the lives of children during that period. In the land rich, but severely labor-scarce economy of early Anglo-America, children by necessity, if not by design, had to contribute to the economy as producers, unlike the children today who function primarily as consumers.[25] For a couple to remain childless in early America was not viewed as an avenue to prosperity, but rather as an abnormal condition associated with impotence, illness, and slavery. Early American adults, including slaves, were no doubt encouraged to have children in part because of their economic importance. Thus, birth rates remained relatively high by modern standards throughout the colonial period.[26]

The effects of children's economic value aside, one of the most controversial questions raised by demographic research concerns the potential impact of the high infant mortality rates on the emotional bonding of early American parents with their children. Some historians have argued that the likelihood of losing children was so great during this period that parents were reluctant to make a full emotional investment in a newborn child until they thought that he or she had a reasonable chance of surviving.[27] These historians assert further that the high birth rates combined with the high death rates encouraged parents to view children as interchangeable, in that a dead infant could in some respects be "replaced" by the birth of another child. Other historians, such as Peter Slater, have challenged this interpretation, and argued instead that early American parents were strongly bonded with their children, but that when faced with a seriously ill child, these parents usually engaged in anticipatory mourning as a way of preparing themselves emotionally for the impact of loss when and if it came, which of course it did with depressing regularity.[28] Thus, Cotton Mather, an eighteenth-century Boston minister who suffered the tragic loss of fourteen of his sixteen children, said the loss of a child was like having a limb torn from one's body.[29]

How these striking conditions actually affected the subjective experience of early American children themselves is difficult to ascertain. Their lives were filled with instability, unpredictability, and loss. To watch one's mother or father, brother or sister, or other close relative or friend die slowly, painfully, without the benefit of pain-killing drugs, was undoubtedly traumatic for young children. If children today mourn the separation of parents through divorce, how did early American children respond when the ravages of an epidemic took their loved ones from them forever? Children in early America had to come to terms with death and hardship in ways that our children and we can hardly imagine. It is both inspiring and mysterious that early American children had the strength and courage to adjust to these conditions.[30]

CHILDREN IN NINETEENTH-CENTURY AMERICA

Historical scholarship on the lives of children in nineteenth-century America produces a mixed picture of change and continuity, but on one point there is widespread agreement among scholars. A profound change in adult attitudes toward children occurred during this period. What began in the late eighteenth and early nineteenth centuries as a gradual weakening of the Calvinist view of the child as innately depraved had been transformed by 1900, at least among the educated classes, into the modern cult of the child.[31] Children were increasingly romanticized and sentimentalized. This process can be documented in poetry, fiction, art, children's literature, child-rearing manuals, and in popular literature. As Bernard Wishy has so aptly put it, "the child redeemable" became "the child redeemer."[32]

Paradoxically, this more positive view of children's character gained strength when American parents began to have significantly fewer children. The average number of children borne by women decreased from 7.4 for those married in the eighteenth century, to 4.9 for those married between 1800 and 1849, and 2.8 for women who were married between 1870 and 1879.[33] Today, the average number of children has dropped below two per couple. Some have argued that with fewer children to care for, parents had more time and emotional energy to invest in each child.[34]

The growing sentimentalization of children during the first half of the nineteenth century was paralleled by, and closely linked to, a powerful movement to idealize women and confine them—especially those from the middle class—to a woman's sphere, which was defined essentially as the home and closely connected activities or environments. By the middle of the nineteenth century, a full-blown ideology was in place that assumed "respectable" women would concentrate their activities on homemaking and child nurture. On the one hand, this clearly focused more attention and human resources on child rearing, but it also encouraged an intensification of interest in and promotion of gender differences among young children. The life courses of boys and girls in the nineteenth century diverged at a very early age and remained sharply differentiated thereafter.[35]

The importance of the greater emphasis on gender is evident, but to what extent did the shift toward a generally more positive, if highly differentiated, image of children make any real difference in the way they were actually treated, or significantly modify the conditions in which they lived? One rather surprising effect of the increasingly powerful interlocking system of ideas concerning children, women, families, and child nurture was to produce a demand to raise the age for school entry. During the colonial period it was not uncommon in New England for children as young as four or even three years of age to attend dame or primary schools. This pattern of early school attendance was encouraged in the 1820s by the infant school movement that spread from Europe to America. Not unlike the modern Head Start program, the first infant schools in America were designed primarily for poor children. However, middle and upper class parents soon awakened to the possible benefits of early education for their children, and these schools became very popular. Yet, as the importance of mothers in early child nurture began to be emphasized during the late 1820s and 1830s, serious concerns were raised about the potential harm that early education could do to children, especially those under five or six years of age. Consistent with the emerging ideology of domesticity, critics of early schooling argued strongly that the best possible education for young children was that provided by a loving mother in a secure and stable home. By the 1850s, few children under the age of five were enrolled in school.[36]

It is also clear from private documents that many middle- and upper-class parents began to pay more attention to their children's uniqueness as individuals. We know that the necronymic naming patterns of the eighteenth century had virtually disappeared by 1800, and by 1850 the use of middle names had become wide-spread.[37] Bronson Alcott, a leading educational reformer and child-care theorist, reflects one of the most striking examples of this new awareness. He kept detailed infant diaries in which he carefully recorded the behavior of his own children, as a part of his effort to learn more about child development. In these diaries, Alcott's young children appear almost as demigods, a far cry from the redeemable but innately sinful Puritan children of a century before.[38] Even the conservative Francis Wayland, Baptist minister and highly respected President of Brown University, justified his decision to starve his fif-teen-month old son Heman into submission as a calculated response to the specific behaviors and temperament of this particularly stub-born child. Perhaps to the surprise of modern child rearing experts, Heman developed and maintained a very close relationship with his father and followed in his father's footsteps to become a successful minister and college president.[39]

Throughout the nineteenth century, there were also recurring campaigns against corporal punishment in many institutions, includ-ing families and schools. Horace Mann—the noted Massachusetts educator who favored greater use of female teachers in elementary schools—was a leader in the fight against corporal punishment in homes and schools. Lyman Cobb, a popular author of children's literature, was one of the first public figures to link the physical abuse of children to alcoholism and wife beating. Cobb also asserted that when a male child saw girls beaten by parents or teachers, he would be more likely to beat his own wife and children. Of course, we know that corporal punishment was not eliminated then or now, but the widely publicized campaign against it in the first half of the nineteenth century reflected a significant shift in opinion that increasingly placed the burden of proof on those who believed that beating children was a virtuous social practice.[40]

However, there is considerable evidence that the more positive attitudes toward children and their welfare that appeared during the first half of the nineteenth century too often did not extend beyond the circle of one's own family, neighborhood, class, or ethnic group. During much of the nineteenth century, the actual conditions in which a majority of young children lived were not fundamentally better than those in the eighteenth century. Infant mortality rates remained very high and did not drop significantly before the early twentieth century, when public health measures—such as sanitation programs and the pasteurization of milk—were instituted in our large cities.[41] Furthermore, the close, reciprocal relationships among family, school, and church that had existed in many colonial communities began to weaken, especially after the 1830s. Rapid population growth, immigration, migration, the growth of the factory system, and the appearance of child labor among the poorer classes combined to make much of nineteenth-century America a very chaotic place.[42] As Rowland Berthoff has observed, Americans during this period were "an unsettled people."[43] The most obvious implication of these unstable conditions for young children is that there was less consistent monitoring and surveillance of their welfare than had been present in the small, intrusive communities of the colonial period. Social control was undermined; prohibitions and taboos were weakened; and deviant behavior, including child abuse and neglect, became more possible if not more prevalent.[44]

Not surprisingly, the disorder and uncertainty of nineteenth-century life brought a response that had important consequences for children. During the last half of the century, there was a powerful social movement to create substitutes for the relatively more ordered, integrated patterns of the past that had provided both adults and children some sense of security and acceptance.[45] Illustrative of this movement was the extraordinary public reaction to the experience of Mary Ellen Wilson, a child in New York City who was physically abused by her stepparents. In 1874 her case was taken up by Eldridge Gerry, counsel for the American Society for the Prevention of Cruelty to Animals. After the case came to court and the shocking details of Mary Ellen's mistreatment was revealed, a public sensation was created that sparked a massive national campaign against child abuse, resulting in the formation of scores of societies for the prevention of cruelty to children and the enactment of legislation in many states to protect children from abuse by placing then in institutions or foster homes.[46]

This campaign against child abuse is merely one example of how efforts to protect children in the nineteenth century often led to greater institutionalization of children. The creation of compulsory public schools, kindergartens, Sunday schools, juvenile courts, orphanages, juvenile homes, children's hospitals, public health and nutrition programs, all demonstrate that the state and other community institutions were being asked to take more direct responsibility for children.[47]

This remarkable crusade to intervene in the lives of children has been assessed by historians in very different ways. Some historians have viewed this activity as a humanitarian effort by reformers to expand the circle of concern to include other people's children, as well as their own. They saw the state as the primary instrument for improving the lives of children because the traditional institutions of family and church often proved inadequate in the new industrial America. The work of John Dewey, Judge Ben Lindsey, Jane Addams, Jacob Riis, and others are often cited to illustrate this process.[48]

Other historians have evaluated these activities quite differently. They see nineteenth-century reformers as representative of upper and middle class Americans who felt threatened by the profound social upheavals of the nineteenth century. Thus, according to this view, the reformers—consciously or unconsciously—took action essentially to protect their own economic, political, and cultural interests. Viewed in this light, the proposed reforms and the institutions used to carry them out were fundamentally instruments of social control and the real welfare of young children became a secondary, not a primary concern. These historians assert that we should not be surprised to learn that poverty continued as a serious problem for many children, and that the state itself too often became

an abusive "parent" by forcing children into institutions that were
often more dangerous in the long run than the homes from which
they were allegedly rescued.[49]

Given this apparently contradictory assessment of children's welfare in nineteenth-century America, we should perhaps conclude
that real improvement in the lives of young children occurred at a
much faster rate among some groups of children than others. Children living in small, stable, well integrated communities or in urban
middle- and upper- class families had a much greater opportunity to
experience the benefits of the more positive attitudes toward children that were characteristic of the nineteenth century. Children of
the urban and rural poor were in a much less favorable position.[50]
Then, as now, poverty wielded an enormous influence on the lives of
young children caught in its grasp. Then, as now, those who tried to
help risked both failure and misunderstanding.

CHILDREN IN THE TWENTIETH CENTURY

When we turn our attention to the twentieth century, it seems clear
that the cult of the child that began in the nineteenth century
reached its maturity in the twentieth. The essential innocence of
young children became an implicit, rarely questioned assumption.
Many middle class parents came to identify so closely with this concept of children that in some ways they judged their own worth as
persons in large measure by their successes or failures as parents.[51]
An entire industry of child development and child care experts
developed over the course of this century to assist parents as they
anxiously approached the daunting task of parenting.[52]

Twentieth-century American couples generally had fewer children than ever before, but it would be unwarranted to conclude that
they valued children less than did parents of earlier times.[53] Indeed,
one need only observe the great sense of loss and deprivation experienced by infertile couples today to realize that many adults
remained deeply committed to having children.[54]

The twentieth century also saw great strides in improving the
basic living conditions of most (if not all) American children. In
many respects, the last quarter of the twentieth century can be
viewed as a golden age for children, compared with the widespread
suffering and deprivation of the past.[55] Health care and nutrition

improved substantially.[56] In 1990, a fetus that survived the period from conception to birth, during which it had an approximately one in four chance of being aborted, found its probability of dying before age one was less than ten per thousand live births.[57] (At the beginning of the century, the infant mortality rate was more than one hundred per thousand live births.[58]) Many childhood diseases that ravaged families and children in earlier periods were eradicated or controlled. Children in the late twentieth century suffered far fewer life-threatening illnesses, seldom experienced the loss of a parent or sibling, and were much more likely to have living grandparents or even great grandparents than did children in earlier periods.[59] Child labor was greatly reduced, though not eliminated, in the twentieth century, and schooling became available to almost all children, including children with special needs, many who would were typically denied such opportunity until recent times.[60] Educational opportunities for girls, and for minority children, were significantly improved, especially during the last half of the century when demands for civil rights and social justice became difficult if not impossible to ignore.[61] In many ways, the promise of the nineteenth century was realized. From this perspective, the twentieth century could rightly be called the century of the child.

Even so, there were some very disconcerting trends in the twentieth century that did not necessarily compare favorably with those of earlier centuries, and did not bode well for the future of America's children.[62] During the twentieth century, the lives of American children were affected by a worldwide economic depression, five major military conflicts, including two world wars, and the cold war with its threat of nuclear holocaust. Virtually no American child in the twentieth century escaped the influence of at least one of these events.[63] Although child labor diminished over the century, children's roles as consumers became more important, and childhood increasingly became commodified, which in turn contributed to the emergence of an often-unhealthy consumer culture for children.[64] Furthermore, many of the benefits of modern childhood were obviously not distributed evenly among America's children. Children in poor families were much less likely than those of the more affluent to get the health care, nutrition, housing, day care, and education that they needed.[65] Indeed, Alberto Palloni has concluded that child health is a powerful but often overlooked factor contributing to "intergenerational transmission of inequalities."[66] Thus, it is significant and revealing that in 1985 more than 20 percent of America's children lived in households below the poverty line, which reflected in part the presence of large numbers of both divorced and never-married single parents. Approximately one in four children under eighteen lived in a single-parent household, and over half of mothers with children under six were employed outside the home at a time when the availability of quality day care was closely tied to ability to pay.[67] Maltreatment of children continued to be an extremely serious problem, with the numbers of reported cases of physical and sexual abuse rising to astonishing levels. Child pornography was still present in spite of serious efforts by legislative and law enforcement groups to eradicate it.[68]

In view of these conditions, it would be foolhardy to make the unqualified assertion that life was altogether better for young children in the late twentieth century than before. To be sure, twentieth-century children were generally healthier and better educated, experienced less strict discipline, entered the work force later than yesterday's children, and were probably subjected to less intrusive social control than were children in earlier periods. Yet, can we be certain that the blended families created by divorce or remarriage in 2000 were more nurturing than the reconstituted families produced by parental death and remarriage in the seventeenth-century Chesapeake? Were the young, affluent parents of the late twentieth century necessarily more nurturing caregivers than seventeenth-century parents Anne Bradstreet or Cotton Mather? Did the latchkey children of the 1990s feel more secure and accepted than the carefully monitored and supervised children of the eighteenth and nineteenth centuries? Did late twentieth-century homes with one parent, a brother and/or sister, and a television set necessarily create a more stimulating, healthy environment for young children than the two-parent, sibling-filled households of early America? Were relatively fewer children abused or neglected in 2000 than in 1900? Were minority children in 2000 able to escape the insidious effects of living in a racist society? At the very least, it seems that any sense of progress, though justified in important respects, must be tempered by a realistic recognition that in spite of a growing understanding of child development—and enormous human, technical, and economic resources—twentieth-century Americans, like those in earlier generations, failed to provide all of their children with the care and nurture they needed and deserved.[69]

CONCLUSION

The history of children provides compelling stories that deserve continuing study by scholars in a wide range of fields, including the history of education. For historians of education, one of the first tasks should be to (1) systematically review what has been learned about the history of children in each period and assess its particular implications for understanding the history of education, (2) explore how utilizing explicit educational perspectives can add analytical strength to the historical study of children, and (3) identify areas and topics for potential research that emerge from this process. The history of children and the history of education, though not synonymous, are closely related and would benefit greatly from closer collaboration. At the very least, scholars in these fields should make every effort to ensure that their research and teaching are informed by the best insights from both fields.

Reviewing historical scholarship on children for any period of American history will reveal many promising topics for investigation, but the last half of the twentieth century is especially suggestive. Some of the developments affecting children during this period were virtually unprecedented, and merit careful scrutiny both for their own sake and for the insights they can provide historians of education. Among the many possible areas for research on children in the last half of the twentieth century are the following: children and the cold war, children's health, special populations and access to schooling, civil rights and children's rights, children and the economy, technology and the transformation of children's experiential world, children and advertising, children and the "new" immigration, peer culture, children and postwar feminism, single parent families, divorce, stepchildren and reconstituted families, children and social policy, children and sexuality, ethnicity and childhood, children's poverty in an age of abundance, child abuse, the evolving ideology of childhood, and how children influenced adults and each other. Clearly, the twentieth century brought profound changes to American children, changes that need our continued investigation and careful reflection.

In 1919, American physician Henry Dwight Chapin attended the First International Conference on Child Welfare, organized in the wake of the unprecedented death and destruction brought by World War I. Chapin reported that the conference participants expressed a renewed interest in finding solutions to the "fundamental problem," of identifying new "means of conserving and developing the child life of the world." There was, he wrote, a heightened appreciation that "child life is infinitely precious, and a new determination that in the future there must be permitted no careless warping of child life through poverty or ignorance or selfishness."[70] "Children," he explained,

> give more than they take. They are the great civilizers and humanizers of the race . . . The nurture and care of children . . . constitute the great educators, and the development of character in parents. Without the unconscious but beneficent influence of children we would soon lapse into a possibly refined but selfish barbarism . . . The child is the sphinx of the world, the constant riddle and mystery, before whom all plans of philosophy, codes of ethics, and systems of theology must somehow prove their value . . . It is the feeble hand of a little child which will knock into oblivion any civilization which has porcelain bathtubs and every modern convenience, but makes no place for real homes and children . . . The child is the judge of civilization, for in the last analysis, any community or state may be safely adjudged by the manner in which it approaches the problems of child life.[71]

Chapin's passionate rhetoric may seem somewhat excessive to our twenty-first-century ears. Children may not be, as he implies, the measure of all things. Even so, Chapin's hope that children would not continue to have their precious lives warped by "poverty, ignorance, or selfishness" was indeed a worthy, even noble sentiment. Alas, it is a profound and discomforting commentary on his century and ours that this hope remains unrealized today.

NOTES

1. See Society for the History of Children and Youth (http://www.h-net.org/~child/SHCY/index.htm).
2. Paula Fass is also the Editor in Chief of the three-volume *Encyclopedia of Children and Childhood in History and Society* (New York: Macmillan, 2004). For the most comprehensive collection of original documents on the history of children and youth, see Robert H. Bremner, ed., *Children and Youth in America: A Documentary History*, 3 vols. (Cambridge, MA: Harvard University Press, 1970–74). This collection is being made available online; see Society for the History of Childrren and Youth (http://www.h-net.org/~child/SHCY/useful_links.htm). For an excellent review of recent scholarship, see Julia Grant, "Children Versus Childhood: Writing Children

Into the Historical Record," *History of Education Quarterly* 45 (Fall 2005):
468–90. For reviews of the historical scholarship on children and youth that
appeared in the 1970s and 1980s, see N. Ray Hiner, "The Child in American Historiography: Accomplishments and Prospects," *The Psychohistory Review* vol. 7. (Summer 1978): 13–23; Barbara Finklestein, "Incorporating Children into the History of
Education," *Journal of Educational Thought* 19 (April 1984): 21–41; Bruce Belling-
ham, "The History of Childhood Since the 'Invention of Childhood': Some Issues of
the Eighties," *Journal of Family History* 13 (November 1988): 347–58; Peter
Petschauer, "The Childrearing Modes in Flux: An Historian's Reflections," *Journal
of Psychohistory* (Summer 1989): 3–15; and Hugh Cunningham, "Histories of Childhood," *American Historical Review* 103 (October 1998): 1195–1208

3. Among those recognized are: Stephen Mintz (Merle Curti Award) for *Huck's Raft: A
History Of American Childhood* (Cambridge, MA: Belknap Press of Harvard Univer-
sity Press, 2004); Elliott West (Caroline Bancroft Award) for *Growing Up With The
Country: Childhood On The Far-Western Frontier* (Albuquerque: University of New
Mexico Press, 1989); Wilma King (Outstanding Book Award) for *Stolen Childhood:
Slave Youth In Nineteenth-Century America* (Bloomington: Indiana University Press,
1995); Marie Jenkings Schwartz (Julia Cherry Spruill Award) for *Born in Bondage:
Growing Up Enslaved in the Antebellum South* (Cambridge, MA: Harvard University
Press, 2000); James Marten (Choice Outstanding Book Award) for *The Children's
Civil War* (Chapel Hill: University of North Carolina Press, 1998); Michael Gross-
berg (Littleton-Griswold Award) for *Governing The Hearth: Law And The Family In
Nineteenth-Century America* (Chapel Hill: University of North Carolina Press,
1985); Leroy Ashby (Choice Outstanding Book Award) for *Endangered Children:
Dependency, Neglect, And Abuse In American History* (New York: Twayne, 1997);
William Tuttle (*New York Times* Outstanding Book) for *"Daddy's Gone to War": The
Second WorldWar in the Lives of America's Children* (New York: Oxford University
Press, 1993); and Linda Gordon (Bancroft and Beveridge Awards) for *The Great
Orphan Abduction* (Cambridge, MA: Harvard University Press, 1999).

4. For example, see the Summer 2005 issue of the *Journal of Social History*, which is
devoted to a consideration of "Globalization and Childhood." See also Peter
Stearns's new survey, *Childhood in World History* (New York: Routledge, 2006), and
Paula. Fass, *Children of a New World: Culture, Society, and Globalization* (New York:
New York University Press, 2006).

5. Paul Fatout, ed., *Mark Twain Speaking* (Iowa City: University of Iowa Press, 1976),
131.

6. For a notable exception to this generalization, see Barbara Beatty's *Preschool Education in America: The Culture of Young Children From the Colonial Era to the Present*
(New Haven: Yale University Press, 1995).

7. James Axtell, *The School Upon a Hill: Education and Society in Colonial New England*
(New Haven: Yale University Press, 1974), 95.

8. Barbara Finkelstein, ed., *Regulated Children/Liberated Children: Education in Psy-
chohistorical Perspective* (New York: Psychohistory Press, 1979), 1–2.

9. See the following two recent historiographical essays: Ellen Lagemann, "Does History Matter in Education Research? A Brief for the Humanities in the Age of Science," *Harvard Education Review* 75 (Spring: 2005): 9–24; and Rubin Donato and
Marvin Lazerson, "New Directions in America Educational History: Problems and
Prospects," *Educational Researcher* 29 (November 2000): 4–15. These essays are
very insightful and well written, but give little explicit attention to children.

10. Henry F. Dobyns, *Their Number Became Thinned: Native American Population
Dynamics in Eastern North America* (Knoxville: University of Tennessee Press,

1983), 34; John Daniels, "The Indian Population of North America in 1492," *William and Mary Quarterly* 49 (April 1992): 298–320; Russell Thornton, "Aboriginal North American Population and the Rates of Decline, ca. A.D. 1500–1900," *Current Anthropology* 38 (April 1997): 310–15. For a somewhat different perspective that considers factors in addition to disease that contributed to population decline, see Massismo Livi-Bacci, "The Depopulation of Hispanic America After the Conquest," *Population and Development Review* 32 (June 2006): 199–233.

11. Larry Jones et al., "Environmental Imperatives Reconsidered: Demographic Crisis in Western North America During the Medieval Climatic Anomaly," *Current Anthropology* 40 (April 1999): 137–70; Robert Jackson, *Indian Population Decline: The Missions of Northwestern New Spain, 1687–1840* (Albuquerque: University of New Mexico Press,1993), and Ramon A. Gutierrez, *When Jesus Came the Corn Mothers Went Away: Marriage and Sexuality in New Mexico* (Stanford: Stanford University Press, 1991).

12. For descriptions of this disaster, see Dobyns, *Their Number Became Thinned*; Jackson, *Population Decline*; David Stannard, "Disease and Infertility: A New Look at the Demographic Collapse of Native Populations in the wake of Western Contact," *Journal of American Studies* 24 (December 1990): 325–50; Dobyns, "Puebloan Historic Demographic Trends," *Ethnohistory* 49 (Winter 2002): 170–204; and Livi-Bacci, "The Depopulation of Hispanic America After the Conquest."

13. Margaret C. Szasz, "Native American Children," in *American Childhood: A Research Guide and Historical Handbook*, ed. Joseph M. Hawes and N. Ray Hiner (Westport, CT: Greenwood, 1985), 311–42; and Gutierrez, *When Jesus Came the Corn Mothers Went Away*. For an insightful study of how whites interpreted epidemics among Native Americans, see David S. Jones, *Rationalizing Epidemics: Meanings and Uses of American Indian Mortality Since 1600* (Cambridge, MA: Harvard University Press, 2004).

14. Jim Potter, "Demographic Development and Family Structure," in *Colonial British America: Essay in the New History of the Early Modern Era*, ed. John Greene and J. R. Pole (Baltimore: Johns Hopkins University Press, 1984), 123–56; and Ross W. Beales, "The Child in Seventeenth-Century America," in *American Childhood: A Research Guide and Historical Handbook, ed. Joseph M. Hawes and N. Ray Hiner* (Westport, CT: Greenwood, 1985), 3–56; see also Douglas H. Ubelaker, "Patterns of Disease in Early North American Population," in *A Population History of North America*, ed. Michael R. Haines and Richard H. Steckel (Cambridge, MA: Cambridge University Press, 2000), 51–98; Henry A. Gemery, "The White Population of the Colonial United States, 1670–1790," in *A Population History of North America*; and Robert Wells, *Revolutions in American's Lives: A Demographic Perspective on the History of Americans, Their Families, and Their Society* (Westport, CT: Greenwood, 1982).

15. Ross Beales, "Child in Seventeenth-Century America," 19.

16. Ibid., 20.

17. Daniel M. Scott and Bernard Wishy, eds., *America's Families: A Documentary History* (New York: Harper and Row, 1982), 4.

18. Darrett B. Rutman and Anita H. Rutman, "Now Wives and Sons-in-Law: Parental Death in a Seventeenth-Century Virginia County," in *The Chesapeake in the Seventeenth Century* (New York: Norton, 1979), 173.

19. Beales, "The Child in Seventeenth-Century America," in *American Childhood*, 3–56, and Constance B Schulz, "Children and Childhood in the Eighteenth Century, in *American Childhood*, 57–109.

20. Cited in Schulz, "Children and Childhood in the Eighteenth Century," 68.

21. Ubelaker, "Patterns of Disease"; Gemery, "White Population of the Colonial United States"; Schulz, "Children and Childhood in the Eighteenth Century"; and Jim Potter, "Demographic Development and Family Structure" in *Colonial British America: Essay in the New History of the Early Modern Era*, ed. John Greene and J. R. Pole (Baltimore: Johns Hopkins University Press, 1984), 123–56

22. Lorena Walsh, "The African American Population of the Colonial United States; Jacqueline Jones, *Labor of Love: Labor of Sorrow: Black Women, Work, and the Family from Slavery to the Present* (New York: Basic Books, 1985); Alan Kulikoff, *Tobacco and Slaves: The Development of Southern Cultures in the Chesapeake, 1680–1800* (Chapel Hill: University of North Carolina Press, 1986); and Jim Potter, "Demographic Development and Family Structure," in *Colonial British America: Essay in the New History of the Early Modern Era*, ed. John Greene and J. R. Pole (Baltimore: Johns Hopkins University Press, 1984), 123–56.

23. Walsh, "African American Population of the Colonial United States"; Russell R. Menard, "The Maryland Slave Population, 1685 to 1730: A Demographic Profile of Blacks in Four Counties," *William and Mary Quarterly* 32 (January 1975): 29–54 and Kulikoff, *Tobacco and Slaves.*

24. For assessments of children in slavery during the antebellum period, see Marie Jenkins Schwartz, *Born in Bondage: Growing Up Enslaved in the Antebellum South* (Cambridge, MA: Harvard University Press, 2001); Wilma King, *Stolen Childhood: Slave Youth in Nineteenth-Century America* (Bloomington: Indiana University Press, 1995); and J. C. Stone, *Slavery Southern Culture, and Education in Little Dixie, Missouri, 1820–1860* (New York: Routledge, 2006). See also the August 2006 issue of *Slavery and Abolition*, which is a special issue devoted to "Children in European Systems of Bondage."

25. Marcus Jernegan, *Laboring and Dependent Classes in Colonial America* (New York: Frederick Ungar, 1931); Viviana A. Zelizer, *Pricing the Priceless Child: The Changing Social Value of Children* (New York: Basic Books, 1985); and Hugh D. Hindman, *Child Labor: An American History* (Armonk, NY: M. E. Sharpe, 2002).

26. Wells, *Revolutions in American's Lives*, 21–68.

27. Nancy S. Dye and Daniel B. Smith, "Mother Love and Infant Death, 1750–1920," *Journal of American History* 73 (September 1986): 329–53; and Daniel B. Smith, "The Study of Families in Early America: Trends, Problems, and Prospects," *William and Mary Quarterly* 19 (January, 1982): 3–28.

28. Peter Slater, "From the Cradle to the Coffin: Parental Bereavement and the Shadow of Infant Damnation in Puritan Society," *The Psychohistory Review* 6 (1977–78): 32–51.

29. Cotton Mather, *Victorina* (Boston: B. Green, 1717). See also Maris A. Vinovskis and Gerald F. Moran, *Religion, Family, and the Life Course* (Ann Arbor: University of Michigan Press, 1992), 109–40, 209–31.

30. For an excellent recent collection of original essays on children in colonial America, See James Marten, ed., *Children in Colonial America* (New York: New York University Press, 2006). Also see Peter Benes, ed., *The Worlds of Children, 1620–1920*, Annual Proceedings of the Dublin Seminar for New England Folklife (Boston: Boston University, 2002).

31. Sterling Fishman, "The Double Vision of Education in the Nineteenth Century: The Romantic and the Grotesque," in *Regulated Children/Liberated Children*, 96–113; Barbara Finkelstein, "Casting Networks of Good Influence: The Reconstruction of Childhood in the United States," in *American Childhood*, 111–52; Jacqueline S. Reinier, *From Virtue to Character: American Childhood, 1775–1850* (New York,

Twayne, 1996), 72–124; and Priscilla S. Clement, *Growing Pains: Children in the Industrial Age, 1850–1890* (New York: Twayne, 1997), 150–85.

32. Bernard Wishy, *The Child and the Republic: The Dawn of Modern American Child Nurture* (Philadelphia: University of Pennsylvania Press, 1968), 1, 7. Also see Gail S. Murray, *American Children's Literature and the Construction of Childhood* (New York: Twayne, 1998), 51–116; Charles Strickland, *Victorian Domesticity: Families in the Life and Art of Louise May Alcott* (Tuscaloosa, AL: University of Alabama Press, 1985); Ann Douglas, *The Feminization of American Culture* (New York: Knopf, 1977); and Judith A. Plotz, "The Perpetual Messiah: Romanticism, Childhood, and the Paradoxes of Human Development," in *Regulated Children/Liberated Children*, 63–95.

33. Wells, *Revolutions in American's Lives*, 92. J. David Hacker argues that the major decline in general American fertility did not occur until after 1840, and that marital fertility remained high until after the Civil War. See J. David Hacker, "Rethinking the 'Early' Decline of Marital Fertility in the United States," *Demography* 40 (November 2003): 605–20. By the late twentieth century, the number of children dropped below two per couple. Also see Gloria L. Main, "Rocking the Cradle: Downsizing the New England Family," *William and Mary Quarterly* 37 (Summer 2006): 35–58.

34. Zelizer, *Pricing the Priceless Child*; Donald H. Parkerson and Jo Ann Parkerson, "Fewer Children of Greater Spiritual Quality: Religion and the Decline of Fertility in Nineteenth-Century American," *Social Science History* 12 (1988): 49–70.

35. Maris Vinovskis, "Family and Schooling in Colonial and Nineteenth-Century America," *Journal of Family History* 12 (1987): 19–37; Anne M. Boylan, "Growing Up Female in Young America," in *American Childhood*, 153–84; Susan B. Norton, "The Evolution of White Woman's Experience in Early America," *American Historical Review* 89 (June 1984): 593–619; Anne M. Boylan, *The Origins of Women's Activism: New York and Boston: 1790–1840* (Chapel Hill: University of North Carolina Press, 2002); Jane Hunter, *How Young Ladies Became Girls: The Victorian Origins of American Girlhood* (New Haven: Yale University Press, 2003); and Julia Grant, "A 'Real Boy' and Not a Sissy: Gender, Childhood, and Masculinity, 1890–1940," *Journal of Social History* 37 (Summer 2004): 829–51.

36. Carl Kaestle and Maris Vinovskis, "From Apron Strings to ABC's: Parents, Children, and Schooling in Nineteenth-Century Massachusetts," in *Turning Points: Historical and Sociological Essays on the Family* (Chicago: University of Chicago Press, 1978), 39–80; Gerald Moran and Maris Vinovskis, "The Great Care of Godly Parents: Early Childhood in Puritan New England"; and Charles Strickland, "Paths Not Taken: Seminal Models of Early Childhood Education in Jacksonian America," in *Handbook of Research in Early Childhood Education*, ed. Bernard Spodek (New York: Free Press, 1982): 321–40.

37. Daniel S. Smith, "Child-Naming Practices, Kinship Ties, and Change in Family Attitudes in Hingham, Massachusetts, 1641–1800," *Journal of Social History* 18 (Summer 1985): 541–66; and J. David Hacker, "Child Naming, Religion, and Decline of Marital Fertility in Nineteenth-Century America," *The History of the Family: An International Journal* 4 (September 1999): 339–65.

38. Charles Strickland, *Victorian Domesticity*, and Charles Strickland, "A Transcendentalist Father: The Child Rearing Practices of Bronson Alcott," *Perspectives in American History* 3 (1969): 5–73.

39. William McLoughlin, "Evangelical Child-Rearing in the Age of Jackson," *Journal of Social History* 9 (1975): 21–43.

40. Myra C. Glenn, *Campaigns Against Corporal Punishment* (Albany: State University of New York Press, 1984); and Elizabeth Pleck, *Domestic Tyranny: The Making of*

American Social Policy Against Family Violence from Colonial Times to the Present (New York: Oxford University Press, 1987).

41. Michael R. Haines, Lee A. Craig, and Thomas Weiss, "The Short and the Dead: Nutrition, Mortality, and the 'Antebellum' Puzzle' in the United States," *Journal of Economic History* 63 (June, 2003): 382–413; Samuel H. Preston and Michael R. Haines, *Fatal Years: Child Mortality in Late Nineteenth-Century America* (Princeton: Princeton University Press, 1991; Richard Meckel, *Save the Babies: American Public Health Reform and the Prevention of Infant Mortality, 1850–1929* (Baltimore: Johns Hopkins University Press, 1990); and Wells, *Revolutions in American's Lives*, 91–208.

42. For example, see Maris Vinovskis' analysis of "The Crisis in Moral Education in Antebellum Massachusetts," in his *Education, Society, and Economic Opportunity: A Historical Perspective on Persistent Issues* (New Haven: Yale University Press, 1995), 45–72; and Timothy J. Gilfoyle, "Street-Rats and Gutter-Snipes: Child Pickpockets and Street Culture in New York City, 1850–1890," *Journal of Social History* 37 (2004): 853–82.

43. Roland Berthoff, *An Unsettled People: Social Order and Disorder in American History* (New York: Harper and Row, 1971).

44. Paul Gilje, "Infant Abandonment in Early Nineteenth-Century New York," *Signs* 8 (1983): 580–90; Priscilla F. Clement, "The City and the Child, 1860–1885," in *American Childhood*, 235–72; Ronald D. Cohen, "Child Saving and Progressivism, 1885–1925,"in *American Childhood*, 273–309; Selma Berrol, *Growing Up American: Immigrant Children in America, Then and Now* (New York: Twayne, 1995); and Elliott West, *Growing Up with the Country: Childhood on the Western Frontier* (Albuquerque: University of New Mexico Press, 1989), 147–78, 213–44.

45. Joseph M. Hawes, *Children in Urban Society: Juvenile Delinquency in Nineteenth-Century America* (New York: Oxford University Press, 1971); LeRoy Ashby, *Saving the Waifs: Reformers and Dependent Children, 1890–1917* (Philadelphia: Temple University Press, 1984); Sonya Michel, *Children's Interests/Mother's Rights: The Shaping of America's Child Care Policies* (New Haven: Yale University Press, 2000); William Carp, ed., *Adoption in America: Historical Perspectives* (Ann Arbor: University of Michigan, 2002); and Lori Askeland, ed., *Children and Youth in Adoption, Orphanages, and Foster Care: A Historical Handbook and Guide* (Westport, CT: Greenwood, 2006)

46. N. Ray Hiner, "Children's Rights, Corporal Punishment, and Child Abuse: Changing American Attitudes, 1870–1920," *Bulletin of the Menninger Clinic* 43 (1979): 233–48; Elizabeth Pleck, *Domestic Tyranny*; Barbara Finkelstein, "A Crucible of Contradictions: Historical Roots of Violence Against Children in the United States," *History of Education Quarterly* 40 (Spring 2000): 1–21; and Michael Grossberg, *Governing the Hearth: Law and the Family in Nineteenth-Century America* (Chapel Hill: University of North Carolina Press, 1985).

47. Carl Kaestle, *Pillars of the Republic: Common Schools and American Society, 1780–1860* (New York: Hill and Wang, 1983); William J. Reese, *America's Public Schools: From the Common School to 'No Child Left Behind'* (Baltimore: Johns Hopkins, 2005); James Marten, *Childhood and Child Welfare in the Progressive Era: A Brief History With Documents* (Boston: Bedford, 2004); Lawrence Cremin, *American Education: The National Experience* (New York: Harper and Row, 1980); Patricia Rooke, "The Child Institutionalized in Canada, Britain, and the United States," *The Journal of Educational Thought* 11 (August 1977): 156–71; Ronald D. Cohen, "Child Saving and Progressivism," in *American Childhood*, 273–309; Joseph M. Hawes, *The Children's Rights Movement: A History of Advocacy and Protection* (Boston: Twayne, 1991); Mary Ann Mason, *From Father's Property to Children's Right: The History of*

182   N. RAY HINER

Child Custody in the United States (New York: Columbia University Press, 1994); and
Richard Meckel, *Save the Babies.*
48. William J. Reese, "The Origins of Progressive Education," *History of Education
Quarterly* 41 (Spring 2001): 1–24; Clement, "The City and the Child"; Cohen,
"Child Saving and Progressivism"; D'Ann Campbell, "Judge Ben Lindsey and the
Juvenile Court Movement," *Arizona and the West* 18 (Spring 1976): 5–20; and
Ashby, *Saving the Waifs.*
49. Clement, "The City and the Child"; Cohen, "Child Saving and Progressivism"; Mary
E. Odem, *Delinquent Daughters: Protesting and Policing Adolescent Female Sexuality
in the United States, 1885–1920* (Chapel Hill: University of North Carolina Press,
1995); Michael Katz, *Reconstructing American Education* (Cambridge, MA: Harvard
University Press, 1987); Dominick Cavallo, "The Politics of Latency: Kindergarten
Pedagogy, 1860–1930," in *Regulated Children/Liberated Children*, 158–83; Barbara
Finkelstein, *Governing the Young: Teacher Behavior in Popular Primary Schools in
Nineteenth-Century United States* (New York: Falmer, 1989); Berrol, *Growing Up
American*; David Wolcott, *Cops and Kids: Policing Juvenile Delinquency in Urban
America, 1890–1940* (Columbus: Ohio State University Press, 2005); and David
Adams, *Education for Extinction: American Indians and Border School Experience,
1875–1928* (Lawrence: University Press of Kansas, 1995).
50. For a perceptive assessment of the lives of children in the rural Midwest, see Pamela
Riney-Kehrberg, *Childhood on the Farm: Work, Play, and Coming of Age in the Midwest* (Lawrence: University Press of Kansas, 2005). In her recent book, Karen
Sanchez-Eppler emphasizes the role children played in creating their own social
meaning: *Dependent States: The Child's Part in Nineteenth-Century American Culture* (Chicago: University of Chicago Press, 2005). Also see Marilyn Holt, *The
Orphan Trains: Placing out in America* (Lincoln: University of Nebraska Press,
1992); Katz, *Reconstructing American Education*; Susan Tiffin, *In Whose Best Interest: Child Welfare Reform in the Progressive Era* (Westport, CT: Greenwood, 1982);
Sherri Broder, *Tramps, Unfit Mothers, and Neglected Children: Negotiating the Family in Nineteenth-Century Philadelphia* (Philadelphia: University of Pennsylvania
Press, 2002); Linda Gordon, *The Great Arizona Orphan Abduction* (Cambridge, MA:
Harvard University Press, 1999); and Steven Lassonde, *Learning to Forget: Schooling
and Family Life in New Haven's Working Class, 1870–1940* (New Haven: Yale University Press, 2005).
51. Gary Cross, *The Cute and the Cool: Wondrous Innocence and Modern American Chil-
dren's Culture* (New York: Oxford University Press, 2004); Wishy, *Child and the
Republic*; and Zelizer, *Pricing the Priceless Child.* Also see Bernard Mergen, *Play and
Playthings: A Reference Guide* (Westport, CT: Greenwood, 1982), and Gary Cross,
Kids' Stuff: Toys and the Changing World of American Childhood (Cambridge, MA:
Harvard University Press, 1997).
52. See especially Barbara Beatty, Emily D. Cahan, and Julia Grant, eds., *When Science
Encounters the Child: Perspectives on Education, Parenting, and Child Welfare in
Twentieth Century America* (New York: Teachers College Press, 2006). Also see Peter
N. Stearns, *Anxious Parents: A History of Modern Child Rearing in America* (New
York: New York University Press, 2004); Julia Grant, *Raising Baby By the Book: The
Education of American Mothers* (New Haven: Yale University Press, 1998); Hamilton
Cravens, *Before Head Start: The Iowa Station & America's Children* (Chapel Hill:
University of North Carolina Press, 1993); Christopher Lasch, *Haven in a Heartless
World: The Family Besieged* (New York: Basic Books, 1977); Charles Strickland and

Andrew M. Ambrose, "The Baby Boom, Prosperity, and the Changing Worlds of Children," in *American Childhood*, 533–85; Nancy P. Weiss, "Mother, the Invention of Necessity: Dr. Spock's *Baby and Child Care*," *American Quarterly* 29 (1977): 519–46; and Murray Levine and Adeline Levine, *Helping Children: A Social History* (New York: Oxford University Press, 1992).

53. Elaine May, *Born in the Promised Land: Childless Americans and the Pursuit of Happiness* (Cambridge, MA: Harvard University Press, 1995); Margaret Walsh and Wanda Ronner, *The Empty Cradle: Infertility in America from Colonial Times to the Present* (Baltimore: Johns Hopkins University Press, 1996); Zelizer, *Pricing the Priceless Child*; and Wells, *Revolutions in American's Lives*, 230–38, 241–62.

54. The large number of induced abortions can obviously be used as counter argument to this view. However, the number of legal abortions in the U.S. peaked in 1990 and slowly declined thereafter. See U.S. Bureau of the Census, *Statistical Abstract of the United States, 2000* (Washington, DC: Government Printing Office, 2000), 78–82. Also see E. Brady Hamilton and Stephanie J. Ventura, "Fertility and Abortion Rates in the United States, 1960–2000," *International Journal of Andrology* 29 (February 2006): 34–45.

55. See especially Section 31, "20th Century Statistics," in *Statistical Abstract of the United States, 2000*, 867–89; and Robert V. Daniel, *The Fourth Revolution: Transformation in American Society from the Sixties to the Present* (New York: Routledge, 2005). Also see Wells, *Revolutions in American's Lives*, 211–85; and John Modell, *Into One's Own: From Youth to Adulthood in the United States, 1920–1975* (Berkeley: University of California Press, 1989).

56. Charles King, *Children's Health in America: A History* (New York: Twayne, 1993); David Cutter, "The Role of Public Health Improvements in Health Advance: The Twentieth-Century United States," *Demography* 42 (February 2005): 1–22; Alexandra M. Stern and Howard Markel, eds., *Formative Years: Children's Health in the United States, 1880–2000* (Ann Arbor: University of Michigan Press, 2002); U.S. Bureau of the Census, *Historical Statistics of the United States From Colonial Times to the Present* (Washington, DC: Government Printing Office, 1975); and *Statistical Abstract of the United States: 2000*, 63–104, 874–79.

57. *Statistical Abstract of the United States, 2000*, 65, 78, 88; and Select Committee on Children, Youth and Families, U.S. House of Representatives, *U.S. Children and Their Families: Current Conditions and Recent Trends* (Washington, DC: Government Printing Office, 1989), 247.

58. *Historical Statistics of the United States*, 26–27.

59. Peter Uhlenberg, "Death and the Family," *Journal of Family History* 5 (1980): 313–20.

60. Walter Trattner, *Crusade for the Children: A History of the National Labor Reform in America* (Chicago: Quadrangle Books, 1970); Hugh Hindman, *Child Labor in American History*; Jacqueline Jones, *A Social History of the Laboring Classes From Colonial Times to the Present* (Malden, MA: Blackwell, 1999); Barry Franklin, *From "Backwardness" to "At Risk": Childhood Learning Difficulties and the Contradictions of School Reform* (Albany: State University of New York Press, 1994); David Nasaw, *Children of the City at Work and at Play* (Garden City: Doubleday, 1985); David I. Macleod, *The Age of the Child: Children in America, 1890–1920* (New York: Twayne, 1998); Joseph M. Hawes, *Children Between the Wars: American Childhood, 1920–1940* (New York: Twayne, 1997); and Elliott West, *Growing Up in the Twentieth Century*, 220, 290–304.

61. *Statistical Abstract of the United States, 2000,* 155–58; Paula S. Fass, *Outside In: Minorities and the Transformation of American Education* (New York: Oxford University Press, 1989); Ruben Donato, *The Other Struggle for Equal Schools: Mexican Americans During the Civil Rights Era* (Albany: State University Press of New York, 1997); James D. Anderson, "The Jubilee Anniversary of *Brown v. Board of Education*: An Essay Review," *History of Education Quarterly* 44 (Spring 2004): 149–57; Victoria M. MacDonald, "Hispanic, Latino, Chicano, or 'Other'?: Deconstructing the Relationship Between Historians and Hispanic Cultural History," *History of Education Quarterly* 41 (Fall 2001): 365–413; and Maris Vinovskis, *The Birth of Head Start: Pre School Education Policies in the Kennedy and Johnson Administrations* (Chicago: University of Chicago Press, 2005).

62. Roberta Wollons, ed., *Children at Risk in America: History, Concepts, and Public Policy* (Albany: State University of New York Press, 1993).

63. Carol K Coburn, *Life at Four Corners: Religion, Gender, and Education in a German-Lutheran Community, 1868–1945* (Lawrence: University Press of Kansas, 1992), 136–46, 148–51; Kristi Lindenmeyer, *The Greatest Generation Grows Up: American Childhood in the 1930s* (Chicago: Ivan Dee, 2005); William M. Tuttle, Jr., *"Daddy's Gone to War": The Second World War in the Lives of American Children* (New York: Oxford University Press, 1993); and Michael Scheibach, *Atomic Narratives and American Youth: Coming of Age with the Atom, 1945–1955* (Jefferson, NC: McFarland, 2003).

64. Lisa Jacobson, *Raising Consumers: Children and the American Mass Market in the Early Twentieth Century* (New York: Columbia University Press, 2005); David T. Cook, *The Commodification of Children, the Children's Clothing Industry and the Rise of the Child Consumer* (Durham, NC: Duke University Press, 2004); and Miriam Formanek-Brunell, *Made to Play House: Dolls and the Commercialization of American Girlhood* (New Haven: Yale University Press, 1993).

65. Linda Gordon, *Pitied But Not Entitled: Single Mothers and the History of Welfare* (New York: Free Press, 1994); Molly Ladd-Taylor, *Mother Work: Women, Child Welfare, and the State, 1890–1930* (Urbana: University of Illinois Press, 1994); Margaret O. Steinfels, *Who's Minding the Children: The History and Politics of Day Care in America* (New York: Simon and Schuster, 1973); LeRoy Ashby, *Endangered Children: Dependency, Neglect, and Abuse in American History* (New York: Twayne, 1997); and John L. Rury, "'Democracy's High School? Social Change and American Secondary Education in the Post-Conant Era," *American Educational Research Journal* 39 (Summer 2002): 307–36.

66. See Alberto Palloni, "Reproducing Inequalities: Luck, Wallets, and Enduring Effects of Childhood health," *Demography* 43 (November 2006): 588–615.

67. Children's Defense Fund, *The State of America's Children* (Washington, DC: Children's Defense Fund, 1991), 143, 148; and *Statistical Abstract of the United States 2000,* 58.

68. Ashby, *Endangered Children,* 150–78; James A. Chu, *Rebuilding Shattered Lives* (New York: John Wiley, 2001); James A. Chu and Elizabeth C. Bowman, eds., *Trauma and Sexuality: The Effects of Childhood Sexual, Physical, and Emotional Abuse on Sexual Identity* (New York: Haworth Medical Press, 2003); Linda Gordon, *Heroes of Their Own Lives: The Politics and History of Family Violence* (New York: Penquin Books, 1988); and Paula S. Fass, *Kidnapped: Child Abduction in America* (New York: Oxford University Press, 1997).

69. Judith Sealander, *The Failed Century of the Child: Governing America's Young in the Twentieth Century* (Cambridge, MA: Cambridge University Press, 2003); and Ashby, *Endangered Children,* 179–185.

70. Henry Dwight Chapin, "The Rights of Childhood," *Good Housekeeping* 69 (December 1919): 39–40.
71. Ibid., 130.

CHAPTER 8

SITES, STUDENTS, SCHOLARSHIP, AND
STRUCTURES: THE HISTORIOGRAPHY OF
AMERICAN HIGHER EDUCATION IN THE
POST-REVISIONIST ERA

Christine A. Ogren

As in the larger field of history of education, revisionism in the
1970s and into the early 1980s profoundly affected the historiography of higher education. The focus of higher education revisionist
critique was traditional accounts of nineteenth-century colleges and
universities, which Marilyn Tobias argued adhered to an "evolution-
ary, linear schema" and presented:

> merely a descriptive chronicle of official actions, usually from the per-
> spective of the college president. These studies suffer from a paucity
> of analysis. Their analysis is usually gratuitous as the process of change
> is neglected, and the relationship between the college and the larger
> community is ignored . . . Moreover, our knowledge and understand-
> ing of higher learning in nineteenth-century America have too fre-
> quently been the result of those who write from the perspective of the
> ascendancy of the university. These retrospective studies view change
> as inevitable, assume a uniform intellectual and social matrix, and see
> university "reformers" as the sole agents of change. Within the dark-
> ness-to-light framework of this historical type, nineteenth-century

colleges have been used as foils to dramatize the directions of the new
universities or have been dismissed as subjects for serious inquiry.

While all historians' work is revisionist to one degree or another, this
wave of scholars—*the* revisionists in the field of history of higher
education—worked to dismiss the "darkness-to-light framework" as
much as to pursue new interpretations; Bruce Kimball criticized
them for being "as much or more concerned to put the traditional
current of higher educational historiography in the wrong as . . . to
do justice to their subject-matter." Revisionist James Axtell detailed
how historians' premature obituaries for the liberal arts college were
"Whig history of the most blatant kind," and David Potts dissected
notions that the antebellum college was narrowly sectarian and its
classical curriculum unpopular. Natalie Naylor disavowed historiography that defined higher education as only colleges and universities,
and Naylor and Colin Burke used new quantitative data to dismantle the long-accepted low attendance figures and high rates of college mortality.[1]

The revisionists' deconstruction of the "darkness-to-light framework" invited historians of higher education to write on a clean slate.
It may seem paradoxical, therefore, that they have consistently continued to look to two *pre*-revisionist founding texts: Frederick
Rudolph's *The American College and University: A History*, first
published in 1962, and Laurence Veysey's *The Emergence of the
American University*, first published in 1965. Rudolph and Veysey
both adhered to the darkness-to-light notion codified earlier by
Richard Hofstader and Walter Metzger. Nevertheless, John Thelin
paid homage to Rudolph's work in an introductory essay to the
1990 reissue of *The American College and University*, and declared
his "debt" to Rudolph in the introduction to his own 2004 *A History of American Higher Education*. In the *History of Education
Quarterly*'s fortieth-anniversary retrospective on Veysey, Christopher Loss testified that his work had "weathered the tests of academic time . . . very well, indeed."[2]

The paradox dissolves with consideration of how Rudolph and
Veysey used the theme of darkness-to-light as well as other aspects of
their scholarship. Rudolph devoted nine of his twenty-two chapters
to various, mostly negative aspects of the antebellum colleges, and
then spent eight chapters on the "new [university] era" that dawned
after the Civil War, mainly extolling its virtues. But Rudolph's style

throughout tends to be more anecdotal than analytical; his narrative account of the development of colleges and universities hangs only loosely on a thin scaffolding of darkness-to-light as he entertains his reader with what Thelin describes as "wit and irony." Veysey begins by referring to the antebellum colleges as "archaic" and their faculty spirit as "gentlemanly amateurism," and ends with a description of the powerful and enduring American university, which he argues was fully formed by 1910. He thus frames his book in darkness-to-light, but this frame is vestigial. Veysey's concern is universities between 1865 and 1910; the outer boundaries fade away as he elucidates four "rival conceptions of higher learning" (discipline and piety, utility, research, and liberal culture) and analyzes of "the price of structure" (how universities reconciled conflicting ideals and ambitions within bureaucratic organization). The core of Rudolph's and Veysey's works could stand without the conceptual framework that revisionists would soon tear down.[3]

Rudolph and Veysey also made important—even revisionist—contributions in spite of their traditional scaffold. At the time Rudolph and Veysey undertook their research, the historiography of higher education suffered from the poor reputation of "house histories," usually hagiographic accounts of the rise and triumph of individual institutions and their leaders. Veysey later lamented this "scholarly tradition that affected only higher education, the custom of aging professors writing celebratory histories of their local campuses. Most such volumes were on the same level as antiquarian local history in general, except that they were written by academics who should have known better." His and Rudolph's books were great strides to move higher education historiography out of the shadow of institutional histories. Rudolph cited information from scores of these volumes but interpreted it through the lens of a prudent historian, demonstrating how these ubiquitous sources could be useful in historical inquiry. Rudolph also took pains in his bibliography to illustrate the depth of existing historical scholarship on higher education aside from institutional histories. Veysey focused on the papers of university leaders and publications of academic associations, and thus largely avoided institutional histories as sources, demonstrating that it was possible to write the history of higher education without them.[4]

While Rudolph's focus on prestigious (by the 1960s) private institutions and state flagship universities mainly in the Northeast,

Midwest, and Pacific West was traditional, he broke new ground in at least three ways. As Thelin points out, his "emphasis on nineteenth-century colleges was relatively novel" because American historians at the time were preoccupied with the colonial period, and his "bold decision to emphasize student life within the history of higher education . . . contributed an analytic model." Rudolph devoted no fewer than three chapters to the extracurriculum, and in 1966 published an article stressing the role of students as agents of college and university change both historically and in the present. He also included a chapter on "The Education of Women"; although he segregated women from the rest of the book and told an uncomplicated story of their "inevitable" appearance in college classrooms, Rudolph did recognize women as actors—albeit marginal ones. Veysey likewise ventured into new historiographical territory. While his focus on a dozen of the best-known universities in the country hardly broke with tradition, he looked at them in a way that challenged barriers between intellectual history and social history, as well as sociology. As Thelin explained, he made the history of institutions "part of the study of organizational behavior"; his account of "the price of structure" was "rich, humorous organizational ethnography." Veysey thus brought university history into the (then new) academic field of higher education and opened a window for the use of sociological approaches to studying the history of university structures.[5]

Post-revisionist historiography of higher education grows from the work of Rudolph and Veysey as well as the demolition efforts of the revisionists, for the former scholars' core contributions survived the collapse of the darkness-to-light myth. The latter group's efforts to liberate the field from tradition also included charting topical gaps and—contrary to Kimball's criticism—carefully mapping new paths of inquiry, often through social-history approaches. Naylor observed that because the "line between [the antebellum] college and academy was blurred," sites of higher schooling such as technical institutes, academies, and female seminaries were a gaping hole in the literature, and Anne Firor Scott began to plug this hole with an account of how one female seminary exposed women to a collegiate curriculum and greatly expanded their horizons. Burke found that normal schools, private business colleges, and other "relatively new institutions" were alternatives to "elitist" postbellum universities, and that antebellum higher education was "student-centered, flexible and highly personal." Potts' exploration of curricula at Baptist and other colleges revealed "steady growth and diversity" as well as students' continuing interest in the classical course of study, which he also pointed out contradicted Rudolph's 1977 book on the history of the curriculum.[6]

Rudolph and Veysey were guilty of overlooking non-collegiate institutions and selling short the scholarly life of the colleges, but the new themes they introduced jibed with other aspects of the revisionist agenda. Rudolph's chapter on women was a halting step beyond a focus only on men, and the revisionists pointed out gender and other gaps in historiographical coverage of who attended higher institutions. Axtell called for more examination of the "social and cultural impact women have made on our colleges and universities, and vice-versa," and Richard Angelo focused on social class and mobility in his discussion of students in Philadelphia in the late nineteenth century. David Allmendinger and Raymond Wolters published books on, respectively, poor students in the early nineteenth century, and African-American students in the 1920s. Their work not only brought underrepresented students into the conversation, but also explored students' role in institutional change. Also following Rudolph's lead, revisionists focused on the nineteenth-century extracurriculum: for example, James McLachlan explained how literary societies enhanced college intellectual life and Joseph Demartini argued that athletics and fraternities offered socialization for success in life. Finally, like Veysey, revisionists were interested in the structures that shaped higher education, but their investigations reached far beyond the research universities. John Whitehead and Jurgen Herbst investigated the states' role in higher education and offered reinterpretations of the impact of the Supreme Court's 1819 Dartmouth College decision on the distinction between public and private. Potts looked into the role of the church in antebellum Baptist colleges and found that denominationalism played a smaller role in establishing and maintaining these institutions than localism, or the efforts of town boosters and local citizens. In addition, Burke pointed to market forces as molders of American collegiate populations and the studies they pursued.[7]

The revisionists' historical explorations, in concert with contributions by Rudolph and Veysey, laid the foundation for a proliferation of new scholarship from the early 1980s into the twenty-first century. The following pages will assess post-revisionist historiography of higher education. The discussion will focus on the century of most interest to the revisionists—the nineteenth—as well as the twentieth century. With more distance from the periods between and following the world wars than Rudolph, Veysey, and the revisionists, post-revisionists have been able to look more closely at these eventful times in higher education history. Post-revisionism also includes some excellent work on the colonial era,[8] but has dealt primarily with the centuries that followed. Four categories capture many of the important topics and issues within this historiography. First, "sites" refers to particular campuses and broader institutional types, which vary greatly in mission and prestige, as well as regional location. Second, "students" deals with who attended, their aspirations, and what they made of their experiences. Third, "scholarship" incorporates the curriculum as well as research. Finally, "structures" includes multiple organizations, both within and outside individual institutions that shape faculty careers and student experiences, institutional mission and resources, and the general contours of higher education. Post-revisionism has made significant contributions to historical understandings of the sites, students, scholarship, and structures of higher education in the United States.

SITES

Most of the two-and-a-half decades worth of research since revisionism has not traveled far from Rudolph and Veysey geographically or in terms of institutional type. Well-known private liberal arts colleges and universities in the Northeast, such as Amherst, Williams, and the Ivy League, remain at the heart of scholarship. Midwestern state flagships such as Indiana University and the universities of Michigan and Wisconsin—as well as the private University of Chicago—are also prominent. The University of California, Berkeley and Stanford University tend to be the token representatives of the West Coast, and southern institutions such as the University of Virginia and Duke University receive the least attention. The institutions featured in Helen Lefkowitz Horowitz's 1987 account of the history of undergraduate campus life are similar to the ones Rudolph highlighted twenty-five years earlier. Roger Geiger's 1986 and 1993 volumes on the history of research universities add only a few campuses to Veysey's short list, and Jerome Karabel's 2005 exposé on the history of college admissions looks at Harvard, Yale, and Princeton. George Marsden explains in the preface to *The Soul of the American University* why he focuses on traditional sites: "Since a fairly limited number of institutions have set the standards for most of the rest of American higher education, I have concentrated on those pace-setting schools." While a small elite has indeed set the pace, the others have varied considerably in their ability—and occasionally in their desire—to keep up; the pace-setters do not tell the complete story. Still, post-revisionist historiography remains for the most part northeastern-bound and, as Linda Eisenmann puts it, "prestige-centric"—with important exceptions.[9]

A few types of post-revisionist monographs make historiographical advances by looking at non-pace-setting sites of higher education. A handful of studies focus on traditional colleges and universities that were not in the vanguard, including institutions in non-traditional regions. The South is the setting for several works published in the 1990s and early 2000s, including Michael Sugrue's article on the formation of southern political philosophy at South Carolina College, Amy Thompson McCandless' and Carolyn Terry Bashaw's books on women students and administrators, Michael Dennis' book on progressivism at southern state universities, and several studies of the history of black institutions or of the desegregation of white colleges and universities. Doris Malkmus and Roger Geiger turn to the rural Midwest, another historiographically non-traditional region. Malkmus finds coeducation thriving in small-town colleges as early as the 1850s, and Geiger uses the example of Ohio to argue that these institutions evolved during the supposed age of the university into "multipurpose colleges." Bruce Leslie interrogates traditional colleges in the most traditional region, the East, but during the late nineteenth and early twentieth centuries, a period when they were no longer in the vanguard. Hugh Davis Graham and Nancy Diamond in *The Rise of American Research Universities: Elites and Challengers in the Postwar Era* examine more than two hundred universities throughout the country, most of which were not prestigious. Graham and Diamond, like the other scholars who venture outside the vanguard and/or into understudied regions, take research on traditional types of institutions in new directions.[10]

Other post-revisionist monographs expand the sites included in the historiography of higher education by turning to less traditional campuses. These include institutions designed to educate a particular group of unempowered or minority students. Barbara Miller Solomon features various women's colleges in her history women and higher education, Horowitz's *Alma Mater* is a cultural and architectural history of the northeastern Seven Sisters Colleges, and Lynn Gordon includes case studies of three women's colleges in her history of progressive-era higher education for women. James Anderson, and Henry Drewry and Humphrey Doermann focus on historically black colleges and universities, while Cary Michael Carney, and Victoria-Maria MacDonald and Teresa Garcia discuss the creation, respectively, of tribal colleges and Hispanic-serving institutions in their histories of higher education for Native Americans and Latinos. Philip Gleason's *Contending with Modernity* presents the history of Catholic institutions in the twentieth century, and Kathleen Mahoney and David Contosta write about Catholic women's colleges, whose students were simultaneously members of two unempowered groups. Less traditional institutions also include those with a vocational or utilitarian orientation. Eldon Johnson and Geiger point out that agricultural, mechanical, and technical schools existed before the Morrill Land-Grant Act of 1862, and that the later land-grant colleges offered education in the liberal arts as well as "useful knowledge." Before evolving into state colleges in the twentieth century, state normal schools had the official function of preparing teachers for the public schools. Jurgen Herbst and Christine Ogren fold these sites into the broader historiography by pointing out that local demand dictated that they also "bring to their community opportunities for advanced education for a variety of purposes" and thus, like agricultural, mechanical, and technical institutions, were not radically different from traditional colleges.[11]

Like normal schools, nineteenth-century academies, seminaries, and high schools lacked the official "college" designation, yet often had the same function; as revisionist Natalie Naylor stated, the "line between college and academy was blurred." Geraldine Joncich Clifford elaborates further: "Many academies, seminaries, high schools, normal schools, and even certain grammar schools offered some 'collegiate' and preprofessional work . . . Given their responsive nature, eclectic curriculum, and the diverse ages of their students, they were college and university annexes at the very least. The common practice of calling the public high schools and normal schools the 'people's colleges' appears fairly accurate on two levels:

academies and high schools were substitutes for college among the poorer or more provincial strata of society, and they were alternatives to college or university proper among Americans in general." Post-revisionist scholarship includes groundbreaking studies of these "people's colleges." In *Chartered Schools*, editors Nancy Beadie and Kim Tolley stress the ubiquitousness of academies throughout the nineteenth century, and the range of topics covered in the dozen plus chapters demonstrates their "responsive nature" and eclectic curricula and students. Beadie uses the notion of "internal improvement" to capture the town boosterism behind the supply of academies, and the students' desire for social and intellectual advancement behind the demand for academies in New York State. In her chapter in *Chartered Schools* and her 2005 book, Margaret Nash presents a rich description of women's education in separate seminaries and coeducational academies during the early national and antebellum eras, demonstrating that these institutions paralleled men's colleges. In addition, David Labaree's *The Making of an American High School* stresses the important role that Philadelphia's Central High School played as a terminal institution for the rising middle class.[12]

While studies of high schools, academies, and normal schools enrich scholarship on the nineteenth century, post-revisionist work on the history of urban universities and junior/community colleges expands significantly the historiography of twentieth-century higher education sites. Featuring these two types of institutions, David Levine's *The American College and the Culture of Aspiration, 1915–1940* is revolutionary. Levine explains the central role that growing urban institutions, which looked and felt different from traditional bucolic campuses, played in expanding the availability of higher education. He presents Atlanta's Emory University as the "prototype" of the urban university, and illustrates the development of municipal or public city institutions with profiles of the University of Akron, Wayne State University, and the College of the City of New York. John Rury explores Chicago's DePaul University as an example of another type of urban institution—the Catholic university that appeared in almost every American city by the early twentieth century—and Fred Beuttler analyzes "controversy over mission" in the establishment and growth of the Chicago Circle Campus of the University of Illinois in the second half of the twentieth century. Levine also observes that "no segment of American higher education expanded so rapidly during the interwar period as the public junior college," the newest "people's college." Stephen Brint and Jerome Karabel present a sociological argument that, throughout their history, junior/community colleges have diverted their students from attending four-year colleges and universities, and Labaree explores the "contradictory mixture of public and private purposes" that has shaped community colleges' ambiguous role as higher education institutions in the twentieth century.[13]

John Thelin refers to the history of different types of higher education institutions as "'vertical history' because they are the familiar landmarks in our institutional consciousness." Although post-revisionist scholarship as a whole remains focused on pace-setting colleges and universities, monographs on non-prestigious and less traditional types of institutions enrich "vertical history." Such works enable Thelin in his recent history of American higher education to pay more attention to historically black institutions and community colleges, and make much more of an attempt to integrate them and women's colleges into the wider story than Rudolph did three-and-a-half decades earlier.[14] Still, "vertical history" is a particularly apt description of post-revisionist historiography of higher education sites—and of Thelin's account—because the subjects of these monographs largely remain in silos, isolated from one another and, more importantly, from traditional, seemingly mainstream, institutions. Allowing these monographs to converse with one another and with histories of better-known institutions will, in turn, advance our understandings of students, scholarship, and structures.

STUDENTS

Post-revisionist scholars have continued the investigations initiated by Rudolph and the revisionists into the extracurriculum and American collegiate populations. Since less traditional schools tended—in some cases purposely, and in others incidentally—to serve particular types of students, monographs on these institutions usually focus to some degree on who attended, their aspirations, and what they made of their experiences. Since the early 1980s, historians of higher education have also asked these questions about the undergraduate students at well-known or prestigious schools. One result is book-length studies of particular activities, from student peace movements in the 1930s, to the history of college athletics, and the evolution of

SITES, STUDENTS, SCHOLARSHIP, AND STRUCTURES

Phi Beta Kappa from a secret debate society to an honor society that occasionally provoked controversy.[15] Another result is more sophisticated understandings of the student bodies at traditional colleges and universities. Histories focusing on gender or women, on particular minority racial, ethnic, or religious groups, or on social class, reveal that unempowered and minority students began to gain access to mainstream institutions in the nineteenth century, and also that access did not ensure equal treatment. Post-revisionist scholarship highlights how variations in students' backgrounds or characteristics have differentiated their college experiences throughout the nineteenth and twentieth centuries. In her history of campus life, Helen Lefkowitz Horowitz illustrates these differences by tracing three "distinct ways of being an undergraduate": college men (and women), outsiders, and rebels.[16] Her work and the work of others on traditional colleges and universities, along with monographs on non-prestigious institutions, greatly expand the historiography of students' characteristics and aspirations, and how they exercised agency to shape their experiences.

The generic students in *The American College and University* were male (and white, and middle-class), but their gender was not a thread in Rudolph's analysis. In arguing that it was the college men who established the first fraternities, and thus "college life," to preserve the spirit of their rebellions against the faculty in the late eighteenth century, Horowitz acknowledges that men were at the center of student culture. At both men's and coeducational colleges, she explains, male (white, middle- and upper-class) students reveled in fraternity life and, beginning in the late nineteenth century, athletics. Kim Townsend is one of the only historians to interrogate the role of gender ideology in men's education, but many post-revisionists focus explicitly on women's experiences, and thus enrich the historiography of gender and higher education. Horowitz describes the first women to attend coeducational and single-sex colleges and universities as outsiders, meaning they were uninterested and/or unwelcome in college life. Barbara Miller Solomon distinguishes between generations of college women; her first generation— in college from the 1860s through the 1880s—corresponds to Horowitz's outsiders, as they were "forthrightly serious; single-minded and conscientious," and "hid neither purposefulness nor anxiety." Still unwelcome in male-dominated activities, some female students in the 1890s and 1900s initiated their own activities, and became "college women" in Horowitz's nomenclature or Solomon's second generation, who "had a more expansive spirit" and "let themselves appear to be at college for the 'pursuit of happiness.'" Beginning in the 1910s, according to Horowitz, college women focused on dating while a third group of students appeared on campus: rebels, including both men and women, scoffed at the insular nature of college life, while concerning themselves with social causes in the wider world. Similarly, Solomon's third generation was "more sophisticated . . . responsive to the Progressive causes of social reform but also flouting conventional mores, she foreshadowed the flapper of the twenties." Lynn Gordon's account of women and higher education in the progressive era uses institutional case studies to add further nuance to understandings of the discrimination college women faced, and the separate culture they created on campus.[17]

While the three generations profiled in such detail by Horowitz, Solomon, and Gordon are at the heart of post-revisionist understandings of female students, other scholars add complexity to the picture of the nineteenth and early twentieth centuries with accounts of women at other types of institutions, and venture further into understudied regions and later decades. Margaret Nash stresses the similarities in academy education for women and men early in the nineteenth century. Christine Ogren's comparison of the women students' experiences at the University of Wisconsin and at the state's normal schools between 1870 and 1920 emphasizes that, while the university women fit into the three generations, "the experiences of normal students inside and outside the classroom did not hinge on gender differences"; these women enjoyed relative intellectual and social equality with male students and even held leadership positions. Turning to the South, Kathryn Walbert argues that the white women who attended academies in the antebellum era "took their educational opportunities quite seriously," while Amy Thompson McCandless explains that gender attitudes and male hostility barred women from, or limited their opportunities at, colleges and universities throughout much of the twentieth century. Focusing on the post-World War II era, Linda Eisenmann illustrates advocacy for women's higher education during a time when "feminism was an identification to be avoided," and other researchers illustrate the bumpy road traveled by the female students who integrated "formerly men's colleges and universities" during the second half of the twentieth century. Being female could thus force a student into out-sider status long after the first generations of women graduated from colleges and universities, but women were relatively free to be at the center of campus intellectual and social life at less traditional institu-tions of the nineteenth century.[18]

Being from a minority racial, ethnic, or religious group was also grounds for outsider status at traditional colleges and universities throughout much of their histories. Horowitz explains that college men and women for many decades "did not become tolerant of those from other ethnic groups, for, although there were a few Jews, Catholics, and Negroes in college, they remained essentially invisi-ble." Other post-revisionists further describe racial discrimination, and also illustrate student agency in the face of difficult circum-stances. The racial group receiving the most attention is African Americans. Cally Waite argues that, beginning in the 1880s, Oberlin College undid decades of integration (which had been radical in the mid-nineteenth century) by gradually segregating its dining halls, dormitories, and literary societies. Linda Perkins documents African American women's struggles to enter the elite Seven Sisters Col-leges; the five hundred who managed to do so before 1960 faced hostility and discrimination in housing, but went on to make impor-tant contributions to the black community. James Anderson dis-cusses how student protesters at black colleges in the 1920s helped to curb some of the racist practices of the institutions' benefactors, and Richard Breaux describes how marginalized black students at midwestern state flagship universities between the World Wars involved themselves in "The New Negro Arts and Letters Move-ment." Studies by Peter Wallenstein and others of court-ordered desegregation in the 1950s and 1960s seek to understand the expe-riences of the first black students at southern universities. In *Black Power on Campus: The University of Illinois, 1965–75*, Joy Ann Williamson analyzes how African American students at one northern university provoked reform during a tumultuous time; they refused to play the role of unobtrusive outsiders.[19]

While participants in the Black Power movement fit Horowitz's definition of rebels, she argues that the first rebels were primarily ethnic/religious minority students, especially Jews, in the early twentieth century. Excluded by the college men and women from fraternity- and sorority-dominated campus life, they refused to be outsiders, seized control of student newspapers and government, and brought societal concerns to campus. While there are no post-revisionist monographs focusing on the experiences of these ethnic/religious pioneers at pace-setting institutions, John Rury mentions the "decidedly ethnic character" of Catholic DePaul University, and David Contosta outlines differences in student life at three twentieth-century Catholic women's colleges in Philadelphia. Not only ethnicity, but students' social-class backgrounds helped to shape these variations in Philadelphia, and the character of DePaul.[20]

Indeed, it is hard to separate social class from ethnicity, religion, or race in students' backgrounds. Not surprisingly, Horowitz explains that most lower-class students who attended prestigious colleges and universities in the nineteenth and twentieth centuries were outsiders in student culture. She reflects: "It is difficult to learn about serious students from conventional sources." Perhaps this is why, as Jana Nidiffer points out, there is a dearth of scholarship on poor students at these institutions. Still, post-revisionist scholarship elucidates how economic status helped to determine which type of institution students attended as well as their experiences once there. Nidiffer and Jeff Bouman also suggest that the numbers of poor students at institutions like the University of Michigan declined beginning in the late nineteenth century. By that time, state normal schools had carved out a niche, albeit unintentionally, as institutions for lower-class students, especially those from rural areas. Jurgen Herbst observes, "We should look to them, not to the land-grant universities, when we speak of the 'democratization of higher education.'" As the majority of normal-school students, unpolished and financially-challenged women and men created a campus life of their own; the students who would be outsiders at traditional colleges and universities dominated the normal schools' very active literary societies, athletics, and clubs—which they used, among other things, as vehicles for attaining the social and cultural capital of the middle class.[21]

David Levine argues that "the culture of aspiration" pervaded higher education by the 1910s, and post-revisionist historiography on less traditional institutions in the nineteenth century suggests that aspiration to attain or maintain middle-class status motivated students to acquire advanced education in earlier decades as well. While normal-school students sought to advance in class status, young people who were already in the middle class sought to cement that status by attending academies or high schools. Nancy Beadie states that "the formation of middle-class networks of culture" was central to "internal improvement" through New York's academies, and Nash asserts that class played a greater role than gender in shaping women's academy education, as advanced education was "part of the formation of a middle-class identity and the consolidation of middle-class cultural imperatives." Furthermore, post-revisionist historiography on traditional institutions demonstrates how "college life essentially confirmed and intensified the growing elitism of late-nineteenth-century American life," as Horowitz states. In *Gentlemen and Scholars*, Bruce Leslie describes in vivid detail how male students at four eastern colleges "developed a more genteel self-image" in the 1860s and 1870s through fraternities, musical groups, and athletics. The "new student life-style" was "a product of rising student wealth. Investing considerable time and money in activities was only feasible for affluent students." By World War I, Leslie explains, "Possibly because the American elite lacked the titles and estates of the British aristocracy that it emulated, colleges became an important source of identity and a primary recipient of elite wealth."[22]

Just as students differentiated themselves by social class within colleges and universities, during the late 1910s and 1920s, Levine explains, "institutions became differentiated in accordance with the socioeconomic backgrounds of the students they attracted." By the end of this period, the "higher the status of an institution, the more it drew its student body from privileged classes in American society." Harold Wechsler and Jerome Karabel outline how the leading colleges and universities developed selective admissions procedures to weed out less-desirable applicants, especially Jews. As the normal schools evolved into state colleges, and the high schools and remaining academies were relegated to secondary rather than higher education in the early twentieth century, the urban universities and junior/community colleges became the new people's colleges. Instead of campus life, historians of these institutions emphasize the commuter nature of their student bodies, which suggests that these aspirants did not (and do not) enjoy access to the lively student culture through which normal-school students aspired to higher-class status. [23]

In the 1930s and beyond, according to Levine, "the democratization of higher education was achieved by the expansion or creation of new types of low-status colleges rather than by democratization of the institutions at the apex of the educational structure." What had taken shape was what Clyde Barrow calls "an expressive myth of universal equal opportunity" masking "an unequal system." Historical studies of expanding access to higher education in the decades following World War II have had to grapple with this myth. In *Soldiers to Citizens*, Suzanne Mettler emphasizes that the "inclusivity" of the Servicemen's Readjustment Act of 1944 (the G.I. Bill) opened doors to higher education for many "from less privileged backgrounds, but also for those who belonged to ethnic, racial, and religious groups that had previously had little hope of such opportunities for education and training." She also acknowledges that many veterans used their G.I. Bill benefits to attend vocational or trade schools, which offered useful job training but decidedly low-status credentials. Sarah Turner and John Bound add that the G.I. Bill did little to improve the higher-educational prospects of African American veterans in the South because segregation barred them from white institutions, and black colleges' resources were too limited to accommodate the many veterans who desired to attend. Horowitz observes that the former soldiers who did attend traditional campuses were serious students who remained outside college life and thus somewhat invisible; and Daniel Clark's analysis of advertisements in the wake the G.I.'s arrival on campus confirms the persistence of traditional colleges' elite image. David Karen argues that in the 1960s, 1970s, and 1980s, while women and blacks narrowed gaps in access even to highly selective colleges and universities, working-class students gained more seats in higher education, but not at the top colleges. Furthermore, the political nature of gains for women and blacks fell into a pattern "that allows for increases at the top but insures that the greatest absolute change is at the bottom of the system. Thus, access to higher education increases; access to elite institutions increases; but the lion's share of the change is concentrated in lowest-tier institutions."[24] In the shadow of the "myth of universal equal opportunity," twentieth-century developments continued trends that had taken shape in the nineteenth century: as diverse types of students gained access to higher education, their social-class background, as well as race/ethnicity and gender, influenced both the institutions they attended and the experiences they and their classmates shaped while there. Institutions and students also differentiated themselves through the curricula offered and pursued, as well as the faculty research they supported.

SCHOLARSHIP

At the center of the "darkness-to-light framework," which the revisionists attacked so vehemently, is the meaning and execution of scholarship—the curriculum and faculty research—in the nineteenth- and early twentieth-century college and university. Just a smattering of post-revisionist monographs focus solely on the undergraduate curriculum. For example, Jurgen Herbst fleshes out revisionist claims that the antebellum college curriculum was actually not out of touch through careful readings of content and context of the 1828 Yale Report, the influential document Rudolph blamed for keeping the colleges in the dark. Herbst argues that the faculty authors of the report in fact displayed flexibility and advocated a balance of languages, literature, and science, rather than a narrowly classical curriculum, and that "by shaping character, forming habits, cultivating taste, and providing opportunities through its scientific lectures to become acquainted with the latest studies and discoveries, a Yale education was designed for leadership in all fields." Furthermore, when Amherst and other colleges ventured further from the traditional curriculum, their students objected. Most post-revisionists discuss the curriculum as part of larger studies. Monographs on academies, normal schools, high schools, and early land-grant colleges all echo Herbst in explaining that their students aspired to study classical literature and other traditional and scientific subjects that conveyed middle-class status. James Anderson, and Henry Drewry and Humphrey Doermann describe the tensions at black institutions in the late nineteenth and early twentieth centuries between African American students who sought a liberal arts education and northern-industrialist benefactors who supported the "Hampton-Tuskegee model" of industrial education. Whether at Yale, Central High School, or Tuskegee Institute, "the people" were interested in a classic liberal arts education.[25]

Investigations of undergraduate education during the rise of the research university, and of democratic tensions in twentieth-century higher education, also contribute to the historiography of the undergraduate curriculum. Roger Geiger explains that during the "era of the multipurpose college" from 1850 to 1890, Yale-Report-inspired denominational colleges in the Midwest began to add elective courses and experiment in professional education, but maintained a classical core. Similarly, Bruce Leslie finds that eastern colleges during the same period incorporated a modified version of the university elective system, maintaining "the notion of a 'unified intellectual experience'"; by the 1910s at these colleges "in 'the age of the university,'" "neither electives nor vocationalism triumphed. Instead, a new consensus emerged based on a definition of liberal education that incorporated breadth, electives, and specialization." Michael Dennis describes a parallel "uneasy equilibrium" in the undergraduate curriculum at southern state universities, and Barbara Miller Solomon explores how the first generations of female students at northern colleges and universities weighed the liberal arts and new gender-specialized fields such as home economics. Catholic institutions struggled during this period to modernize the curriculum, while revering the wisdom from the past; their "compelling critique of modernism," according to Kathleen Mahoney and Caroline Winterer, would become "embedded in the curriculum of Catholic and Protestant higher education in the twentieth century." Meanwhile, as Julie Reuben explains, modernizing prestigious universities with Protestant foundations struggled to maintain a union between knowledge and religious values, ultimately retaining but marginalizing morality in the humanities curriculum and the extracurriculum. Thus, as the nineteenth gave way to the twentieth century, the liberal arts lived on in the universities, and the colleges showed that they could be adaptable and dynamic.[26]

Post-revisionist interrogations of the culture of middle-class aspiration and institutional stratification are instructive regarding curricular issues in the twentieth century. David Labaree encapsulates the core curricular dynamic in his theory of "mutual subversion," through which "the professional has come to dominate the goals of higher education while the liberal has come to dominate its content"—promising both middle-class professional credentials and cultural capital. According to this theory, the oldest private universities and the state flagships constitute the highest tier of institutions because they offer the most liberal curriculum, followed in succession by the land-grant colleges, and then the state colleges and regional universities. Having originated as normal schools, the latter institutions struggled especially strenuously to institute a higher-status curriculum, as they became teachers colleges and then general-purpose institutions during the twentieth century. Relegated to the bottom tier in the hierarchy and to vocational training, the junior/community colleges illustrate how the "expressive myth of universal equal opportunity" played out in the twentieth-century curriculum. As Steven Brint and Jerome Karabel explain, junior- and then community-college students throughout most of the century were most interested in liberal academic subjects that would allow them to transfer to four-year colleges. Nevertheless, beginning in the 1930s, administrators who sought to carve out a niche for the junior colleges developed and promoted vocational and other terminal-education programs. The huge enrollment booms of the post-World War II period drew more attention and funding to these "vocation-alization" efforts, but not until the economic crisis of the 1970s did community college students begin to actively seek vocational pro-grams; by that time, generations of their predecessors had been diverted from pursuing a four-year liberal arts education. Labaree further suggests that a focus on credentials over intellectual growth has diverted students at all tiers of institutions from meaningful learning.[27]

A co-conspirator in diverting attention from meaningful under-graduate learning beginning in the late nineteenth century was the rise of university research. Larry Cuban traces "how scholars trumped teachers" on the Stanford University faculty between 1890 and 1990. At first, he says, an "uneasy compromise" attempted to maintain a traditional undergraduate college within the larger uni-versity structure. "Yet the balance between the teaching imperative buried within the college and the research imperative buried within the graduate school went awry as decades passed. The unrelenting spread of the research-based graduate school culture to educating undergraduates produced far more emphasis on creating mini-aca-demics than on molding citizens." Answering this question became "more difficult as expert knowledge grew and pressures for voca-tional preparation [i.e., credentials] increased." Geiger writes that by the 1960s, due to changes in the faculty role resulting from the rise of research—the "academic revolution" articulated by Christopher Jencks and David Riesman—professors at research universities had lost touch with "the intellectual needs of students . . . the majority of undergraduates suffered from the depreciation of teaching, frag-mented or esoteric courses, and the discouragement by the academic imperium of alternative models." At the behemoth public institutions Clark Kerr termed "multiversities," students' intellectual needs competed not only with faculty research but also with demands that universities serve the greater society. Kerr celebrated the "many groups to be served," but students were often left to fend for themselves. The student protests for which the decade is famous were most widespread at public multiversities and private research universities, and arose in part from students' frustrations with dysfunctional undergraduate education. Twenty years later, Geiger explains, "the source of [students'] alienation was [still] real. Universities had abandoned, largely unconsciously, the moral and cultural stewardship of undergraduates."[28]

The effect of faculty research on undergraduate education is just one dimension of post-revisionist historiography on research universities and the research enterprise. Geiger and other scholars complicate Veysey's vision of "the emergence of the American university" between 1865 and 1910, and describe a second transformation in research institutions in the mid-twentieth century. George Marsden and Julie Reuben question Veysey's notion that champions of science and scientific approaches to research forced religion out of the young universities; both argue that Harvard's Charles W. Eliot, Chicago's William Rainey Harper, and other university builders shared a Christian belief structure *and* a dedication to scientific inquiry. In *To Advance Knowledge*, Geiger traces how sixteen of the leading universities established a research ethos between 1900 and 1940, suggesting that the formation of the research university extended beyond Veysey's endpoint. These institutions took four decades to fully cultivate large enrollments of tuition-paying undergraduates, to recruit and support research-oriented faculty members, and to begin to secure funding from outside sources. Before 1920, the central dynamic was the evolution of a "characteristically American amalgamation of university teaching and research" in which most research was basic science funded on an ad hoc basis, and after 1920, a system took shape in which private philanthropic foundations funded, and sometimes dictated, university research.[29]

Between World War II and the 1960s, new wells of research funding underwrote "academia's golden age" according to Richard Freeland, and the universities' "era of . . . greatest accomplishments" according to Geiger. Freeland explains that the universities' "extensive involvement in the military effort stirred a new awareness of the importance of academic work," which "extended into the postwar period," fostering in research a "tendency to focus on national concerns." Geiger's *Research and Relevant Knowledge* looks at leading universities between 1940 and 1990. He focuses on this period's "research economy," which included not only the federal government but also foundations and corporations; the Ford Foundation is as important to the story as the National Science Foundation. He also follows how universities organized resources and finances, such as through establishing research institutes and federal contract laboratories, and notes the role of campus dialogues and university external relations concerning issues such as the protection of basic science. Rebecca Lowen turns to Stanford during the cold war, arguing that many university leaders and faculty members, "for both institutional and ideological reasons . . . favored and promoted the development of heavily subsidized scientific work." It was because their interests overlapped with those of government and industrial patrons of research, in a "permanent war economy" that "the cold war university" developed as it did. Her analysis helps to explain why, as Ellen Schrecker laments in *No Ivory Tower: McCarthyism and the Universities*, university faculties did not object strenuously to threats to academic freedom during this time. Hugh Davis Graham and Nancy Diamond focus on slightly later periods, comparing "the research performance of faculty" at over two hundred universities through the "'golden years' of the 1960s, the belt-tightening adjustments of the 1970s, the entrepreneurial ethos of the 1980s, and the darkening horizon of the 1990s." They argue that faculty at private universities enjoyed a competitive research advantage over those at public institutions, that the presence of a medical school was a decided advantage for an entire institution in securing grants, and that the late twentieth century nevertheless witnessed the emergence of new elite research institutions, especially within state university systems. These histories of the research enterprise in the twentieth century capture the development of what Ellen Condliffe Lagemann calls the "politics of knowledge"—structures external to the university were reaching far into the ivory tower.[30]

STRUCTURES

Structures internal and external to higher education institutions have helped shape faculty careers and student experiences, institutional mission and resources, and the general contours of higher education throughout its history. In *The Emergence of the American University*, Veysey catalogued "the price of structure" for research universities at the turn of the twentieth century, and post-revisionist historians have elaborated upon the internal administrative issues and vulnerability to corporate influence that were Veysey's main focus. Jana Nidiffer and Timothy Reese Cain examine the role of academic vice presidents/provosts, and Nidiffer and Carolyn Terry Bashaw elucidate how deans of women answered administrative needs and formed their own inter-university structure for professional communication and support. As Philo Hutcheson outlines, faculty also formed an inter-university structure in the 1910s: the American Association of University Professors operated as a professional organization for much of the twentieth century. In his history of college and university governing boards, Edwin Duryea notes the influence of lay trustees, many of whom by the early twentieth century were from the business sector. Hugh Hawkins discusses opposition within the academy to more corporate forms of governance, and asks, "Was the massive industrial economy corrupting academic ideals"? Christopher Newfield's answer is "yes"; in *Ivy and Industry*, he argues that universities in the early twentieth century adopted "divided" governance in which faculty had "freedom to do their own work," but with "built-in costs" including "a preference for bureaucracy over democracy, a permanent vulnerability to business influence, and a weakening of individual agency."[31]

Beyond these largely internal structural issues, post-revisionists have turned to external legislation, foundations, and organizations to understand higher education, particularly since 1900. In a look back at Veysey's book in 1987, John Thelin directed historians' attention to new structures that took shape in the twentieth century: "multicampus systems . . . state coordinating agencies, the largely unforeseen phenomenon of federal research and development grants, and the bureaucratic changes spawned by compliance with government regulation."[32] Historiography on such developments suggests that growing external structures have changed higher education, with a mixed influence on individual and institutional autonomy.

John Aubrey Douglass' study of "the California Idea" illustrates the Golden State's pathfinding role in state coordination of multicampus systems. State leaders during the first two decades of the twentieth century converged on "three interrelated goals": "all high school graduates should have the opportunity for postsecondary training"; "California government should aggressively expand the number of public education institutions"; and, in the process, "new types of institutions and academic programs should be established to cater to the social and economic needs of a rapidly changing California." As a result, he reports, the state played an unprecedented role in structuring public higher education: "By 1920, California government had established a formal and coherent hierarchy of public institutions that could be found in no other state." At the top was the University of California, "the nation's first multicampus university," followed by state teachers colleges that were undergoing a transformation into regional undergraduate liberal arts institutions, and finally "a network of public junior colleges." Douglass stresses "the powerful role of the University of California in creating the tripartite structure," and emphasizes that the revolutionary Master Plan for Higher Education in 1960, which created "a path for ordered growth in the state's higher education system," was "the result of a negotiation process between the higher education community and lawmakers that, in the end, preserved and codified the best aspects of the California Idea."[33]

Just as Douglass recognizes a balance of state control and institutional agency—at least for the flagship university—historians who document the profound effects of growing federal structures in twentieth-century higher education do not present a simple story of government imposition. Central in the work by Roger Geiger, Richard Freeland, Rebecca Lowen, and Hugh Davis Graham and Nancy Diamond on research universities during World War II and the cold war are the federal committees and organizations that established policies on research funding. According to Lowen, as well as Graham and Diamond, federal science policy exacerbated differences between elite and other universities by favoring those which had more resources; when federal science policy-makers by the 1970s added requirements that funds be distributed more widely, they again reshaped the contours of the nation's research universities. Still, peer review of grant applications, awarding grants through contracts, and reimbursement of universities' overhead costs helped to balance power between the academy and the government. Freeland states, "the dominant view—inside and outside of higher education—was that expansion was improving the academy as well as the country." In addition to research, expansion meant increasing significantly the numbers of undergraduate students through additional federal structures, beginning with the 1944 G.I. Bill. According to Robert Serow, this piece of legislation contributed to the profound shift to mass higher education, but only in concert with "complex economic, cultural, and scientific forces." Furthermore, the G.I. Bill was relatively free of federal controls, in part, according to J. M. Stephen Peeps, because higher education leaders worked from the sidelines and through the American Council on Education to assure that the bill recognized individual, institutional, and state power. In her account of the federally-funded educational opportunity and affirmative action programs that began in the 1960s, Julie Reuben notes that universities "gradually reformed minority recruitment programs to bring them in line with selective admission more generally," and in the process limited the programs' scope. These federal programs had important effects on higher education, although they were balanced by universities' assertions of institutional agency.[34]

Like state and federal initiatives, philanthropic structures supported colleges and universities, while posing a potential threat to their autonomy, and post-revisionism includes important research in this area. In his work on early twentieth-century research universities, Geiger describes their "common goals and cross purposes" with foundations. Beginning in the 1920s, philanthropic organizations shaped campus growth by funding the construction of new research buildings and university units, shaped research practice by funding particular approaches such as social surveys, and shaped careers by sponsoring postdoctoral fellowships. But the foundations were not all-powerful, as "fundamental differences in outlook as well as inherent limitations in resources" restricted their reach. Lowen and Ellen Lagemann suggest that philanthropic structures extended their reach in later years, however. During the cold war, according to Lowen, the Ford Foundation was as least partly responsible for the "behavioral revolution" in social science research through the approaches that it favored in grant-giving. And Lagemann entitles her monograph on the history of the Carnegie Corporation *The Politics of Knowledge*, explaining, "The Corporation's self-imposed mandate to define, develop, and distribute knowledge was, in a sense, a franchise to govern, in important indirect ways." Other scholars examine philanthropy in the area of African American higher education; as Marybeth Gassman states, "The history of black colleges is interwoven with that of philanthropy." In *Dangerous Donations*, Eric Anderson and Alfred A. Moss argue that the Rocke-feller-funded General Education Board in the early twentieth century strengthened a small number of black colleges, but stopped short of working to effect change in southern society. According to Gassman, black leader Charles S. Johnson saw such philanthropy "as a pragmatic route with which to create opportunities for African Americans," while W. E. B. Du Bois opposed it out of fear for its potential to control; Johnson focused on the common goals while Du Bois focused on the cross purposes inherent in many philanthropic efforts in higher education.[35]

Finally, the history of nonprofit associations—such as the College Entrance Examination Board (CEEB) and the American Association of University Women (AAUW)—presents many of the same issues as research on government and philanthropic structures. These organizations are present in the historiography of higher education, but tend to be tangential. One exception is *A Faithful Mirror: Reflections on the College Board and Education in America*, edited by Michael Johanek. The one-hundredth anniversary of the CEEB in 2000 provides an opportunity for the authors in this volume to reflect upon the history of student evaluation, financial aid, affirmative action, and relations between colleges and high schools. Harold Wechsler argues that higher education institutions supported the creation of the CEEB in part to diminish what they perceived as the high schools' growing power in their interactions over admissions. A half-century later, the board formed the Educational Testing Service to administer the Scholastic Aptitude Test, with the approval of college officials; "By emphasizing success in college instead of performance in high school, the SAT helped to keep the colleges and high schools at arms length." Thus, colleges and universities sought to gain authority in their relations with the schools though relinquishing some power to an outside organization. Focusing specifically on the Seven Sisters Colleges of the Northeast, Andrea Walton considers how college officials were able to use the structures of agencies including the CEEB and the AAUW to achieve wide objectives. She concludes that membership in these voluntary professional organizations enabled the colleges to collaborate "in efforts to elevate standards in women's higher education . . . They were able to articulate and promote a selective, rigorous vision of single-sex liberal arts collegiate education for women in a world where universities were beginning to dominate and coeducation was the norm." In both of these cases, higher education institutions worked out a power-sharing agreement with potentially life-altering consequences for their constituents. In his history of selective college admissions, Wechsler asks, "How legitimate was and is higher education's regulation of social mobility?" One might reframe the question, how legitimate was and is it for higher education to share authority in the regulation of social mobility? The historiography of higher education elucidates many of the dangers as well as benefits of relationships between external structures and colleges and universities.[36]

CONCLUSION

"The founding and influence of institutions and agencies that cut across the higher education landscape," according to John Thelin, are "horizontal history."[37] While structures are squarely horizontal and individual sites are vertical, students and scholarship can fall into either category—vertical when restricted to one institution or type, and horizontal when studied across types of institutions. Post-revisionists have significantly enriched the literature on the variety of institutional types, on student backgrounds and experiences, on undergraduate learning and faculty research, and on the organizations that have influenced higher education in the nineteenth and twentieth centuries. It is still rare, though, for studies to combine horizontal and vertical history; few monographs discuss students, scholarship, or structures across different sites. Weaving together the vertical and horizontal threads will allow for more systemic studies, as friction between these threads has created formal and informal systems of higher education in the United States. A future review of the historiography of higher education will likely acknowledge the continuing influence of Frederick Rudolph's *The American College and University: A History* and Laurence Veysey's *The Emergence of the American University*, the contributions of the revisionists, and the wealth of post-revisionist scholarship on sites, students, scholarship, and structures. In addition, a future review will hopefully be able to add an additional category: *post* post-revisionist studies of systems of higher education.

NOTES

1. Marilyn Tobias in "Conversation: Renegotiating the Historical Narrative: The Case of American Higher Education," *History of Education Quarterly* 44 (Winter 2004): 583; Marilyn Tobias, *Old Dartmouth on Trial: The Transformation of the Academic Community in Nineteenth-Century America* (New York: New York University Press, 1982), 4; Bruce A. Kimball, "Writing the History of Universities: A New Approach?" *Minerva* 23 (1986): 387, Tobias also quoted on 376; James Axtell, "The Death of the Liberal Arts College," *History of Education Quarterly* 11 (Winter 1971): 341; David B. Potts, "American Colleges in the Nineteenth Century: From Localism to Denominationalism," *History of Education Quarterly* 11 (Winter 1971): 363–80; David B. Potts, "Curriculum and Enrollments: Some Thoughts on Assessing the Popularity of Antebellum Colleges," *History of Higher Education Annual* 1 (1981): 88–109; Natalie A. Naylor, "The Ante-Bellum College Movement: A Reappraisal of Tewksbury's Founding of American Colleges and Universities," *History of Education Quarterly* 13 (Autumn 1973): 261–74; Colin B. Burke, *American Collegiate Populations: A Test of the Traditional View* (New York: New York University Press, 1982).
2. Frederick Rudolph, *The American College and University: A History* (Athens: University of Georgia Press, c. 1962, 1990); Laurence R. Veysey, *The Emergence of the American University* (Chicago: University of Chicago Press, 1965); Richard Hofstadter and Walter P. Metzger, *The Development of Academic Freedom in the United States* (New York: Columbia University Press, 1955); John R. Thelin, "Rudolph Rediscovered," introductory essay to Rudolph, *The American College and University*, ix–xxiii; John R. Thelin, *A History of American Higher Education* (Baltimore: Johns Hopkins University Press, 2004), xix; "Retrospective: Laurence R. Veysey's *The Emergence of the American University*," *History of Education Quarterly* 45 (Fall 2005): 405–60; Christopher P. Loss, "Introduction" to "Retrospective: Laurence R. Veysey's," 405
3. Rudolph, *The American College and University*; Thelin, "Rudolph Rediscovered," xxi; Veysey, *The Emergence*, 2.
4. Veysey, quoted in Lester F. Goodchild and Irene Pancner Huk, "The American College History: A Survey of Its Historiographic Schools and Analytic Approaches from the Mid-Nineteenth Century to the Present," in *Higher Education Handbook of Theory and Research*, vol. 6, ed. John C. Smart (New York: Agathon, 1990), 201; Rudolph, *The American College and University*; Thelin, "Rudolph Rediscovered," xviii–xix; for an in-depth treatment of institutional histories as a genre, see Goodchild and Huk, "The American College History." Institutional histories that defy their poor reputation include: Paul K. Conkin, *Gone with the Ivy: A Biography of Vanderbilt University* (Knoxville: University of Tennessee Press, 1985); Merle Curti and Vernon Carstensen, *The University of Wisconsin: A History* (Madison: University of Wisconsin Press, 1949); David B. Potts, *Wesleyan University, 1831–1910: Collegiate Enterprise in New England* (New Haven, CT: Yale University Press, 1992).
5. Thelin, "Rudolph Rediscovered," xvii, xii–xiii; Rudolph, *The American College and University*, chaps. 5 ("The Collegiate Way"), 7 ("The Extracurriculum"), 18 ("The Rise of Football"), and 15 ("The Education of Women"); Frederick Rudolph, "Neglect of Students as Historical Tradition," in *The College and the Student*, ed.

Lawrence E. Dennis and Joseph F. Kauffman (Washington, DC: American Council on Education, 1966); John R. Thelin, review of *The Emergence of the American University* by Laurence R. Veysey, *History of Education Quarterly* 27 (Winter 1987): 519, 520. Other important works from this time period that crossed boundaries between sociology and history are: Christopher Jencks and David Riesman, *The Academic Revolution* (New York: Doubleday, 1968); E. Alden Dunham, *Colleges of the Forgotten Americans: A Profile of State Universities and Regional Colleges* (New York: McGraw-Hill, 1969).

6. Naylor, "The Ante-Bellum College Movement," 263; Anne Firor Scott, "The Ever-Widening Circle: The Diffusion of Feminist Values from the Troy Female Seminary, 1822–1872," *History of Education Quarterly* 19 (1979): 3–25; Burke, *American Collegiate Populations*, 212, 5; David B. Potts, "'College Enthusiasm!' as Public Response, 1800–1860," *Harvard Educational Review* 47 (February 1977): 38; Potts, "Curriculum and Enrollments." See also Frederick Rudolph, *Curriculum: A History of the American Undergraduate Course of Study Since 1636* (San Francisco: Jossey-Bass, 1977).

7. Axtell, "The Death," 350; Richard Angelo, "The Students at the University of Pennsylvania and the Temple College of Philadelphia, 1873–1906: Some Notes on Schooling, Class and Social Mobility in the Late Nineteenth Century," *History of Education Quarterly* 19 (Summer 1979): 179–205; David F. Allmendinger, *Paupers and Scholars: Transformation of Student Life in Nineteenth-Century New England* (New York: St. Martin's Press, 1975); Raymond Wolters, *The New Negro on Campus: Black College Rebellions of the 1920s* (Princeton: Princeton University Press, 1975); James McLachlan, "The *Choice of Hercules*: American Student Societies in the Early 19th Century," in *The University in Society*, vol. 2, ed. Lawrence Stone (Princeton: Princeton University Press, 1974), 449–94; Joseph R. Demartini, "Student Culture as a Change Agent in American Higher Education: An Illustration From the Nineteenth Century," *Journal of Social History* 9 (Spring 1976): 526–41; John S. Whitehead, *The Separation of College and State: Columbia, Dartmouth, Harvard, and Yale, 1776–1876* (New Haven, CT: Yale University Press, 1973); Jurgen Herbst, "The Eighteenth Century Origins of the Split Between Private and Public in Higher Education in the United States," *History of Education Quarterly* 15 (Fall 1975): 273–79; Jurgen Herbst, *From Crisis to Crisis: American College Government, 1636–1819* (Cambridge, MA: Harvard University Press, 1982); John S. Whitehead and Jurgen Herbst, "Forum: How to Think About the Dartmouth College Case," *History of Education Quarterly* 26 (Fall 1986): 333–49; Potts, "American Colleges," Burke, *American Collegiate Populations*, 262.

8. See, for example, David W. Robson, *Educating Republicans: The College in the Era of the American Revolution, 1750–1800* (Westport, CT: Greenwood, 1985); Bobby Wright, "'For the Children of Infidels'?: American Indian Education in the Colonial Colleges," *American Indian Culture and Research Journal* 12 (1988): 1–14; J. David Hoeveler, *Creating the American Mind: Intellect and Politics in the American Colonial Colleges* (New York: Rowman and Littlefield, 2002); John D. Burton, "The Harvard Tutors: The Beginning of an Academic Profession, 1690–1825," *History of Higher Education Annual* 16 (1996): 5–20; John D. Burton, "Collegiate Living and Cambridge Justice: Regulating the Colonial Harvard Student Community in the Eighteenth Century, " *History of Higher Education Annual* 23 (2004): 83–105.

9. Helen Lefkowitz Horowitz, *Campus Life: Undergraduate Cultures from the End of the Eighteenth Century to the Present* (Chicago: University of Chicago Press, 1987); Roger L. Geiger *To Advance Knowledge: The Growth of American Research Universities, 1900–1940* (New York: Oxford University Press, 1986); Roger L. Geiger,

Research and Relevant Knowledge: American Research Universities Since World War II (New York: Oxford University Press, 1993); Jerome Karabel, *The Chosen: The Hidden History of Admission and Exclusion at Harvard, Yale, and Princeton* (Boston: Houghton Mifflin, 2005); George M. Marsden, *The Soul of the American University: From Protestant Establishment to Established Nonbelief* (New York: Oxford University Press, 1994), vii; Linda Eisenmann, "Essay Review: Women, Higher Education, and Professionalism: Clarifying the View," *Harvard Educational Review* 66 (Winter 1996): 858.

10. Michael Sugrue, "'We Desire Our Future Rulers to be Educated Men': South Carolina College, the Defense of Slavery, and the Development of Secessionist Politics," *History of Higher Education Annual* 14 (1994): 39–71; Amy Thompson McCandless, *The Past in the Present: Women's Higher Education in the Twentieth-Century American South* (Tuscaloosa: University of Alabama Press, 1999); Carolyn Terry Bashaw, *"Stalwart Women": A Historical Analysis of Deans of Women in the South* (New York: Teachers College Press, 1999); Michael Dennis, *Lessons in Progress: State Universities and Progressivism in the New South, 1880–1920* (Urbana: University of Illinois Press, 2001); James D. Anderson, *The Education of Blacks in the South, 1860–1935* (Chapel Hill: University of North Carolina Press, 1988); Henry N. Drewry and Humphrey Doermann, *Stand and Prosper: Private Black Colleges and Their Students* (Princeton, NJ: Princeton University Press, 2001); E. Culpepper Clark, *The Schoolhouse Door: Segregation's Last Stand at the University of Alabama* (New York: Oxford University Press, 1993); Nadine Cohodas, *The Band Played Dixie: Race and the Liberal Conscience at Ole Miss* (New York: Free Press, 1997); Peter Wallenstein, "Black Southerners and Non-Black Universities: Desegregating Higher Education, 1935–1967," *History of Higher Education Annual* 19 (1999): 121–48; Charles S. Padgett, "'Without Hysteria or Unnecessary Disturbance': Desegregation at Spring Hill College, Mobile, Alabama, 1948–1954," *History of Education Quarterly* 41 (Summer 2001): 167–188; Doris Malkmus, "Small Towns, Small Sects, and Coeducation in Midwestern Colleges, 1853–1861," *History of Higher Education Annual* 22 (2002): 33–65; Roger L. Geiger, "The Era of Multi-Purpose Colleges in American Higher Education, 1850–1890," *History of Higher Education Annual* 15 (1995): 51–92; W. Bruce Leslie, *Gentlemen and Scholars: College and Community in the "Age of the University," 1865–1917* (University Park, PA: Pennsylvania State University Press, 1992); Hugh Davis Graham and Nancy Diamond, *The Rise of American Research Universities: Elites and Challengers in the Postwar Era* (Baltimore: Johns Hopkins University Press, 1997).

11. Barbara Miller Solomon, *In the Company of Educated Women: A History of Women and Higher Education in America* (New Haven, CT: Yale University Press, 1985); Helen Lefkowitz Horowitz, *Alma Mater: Design and Experience in the Women's Colleges from Their Nineteenth-Century Beginnings to the 1930s* (New York: Alfred. E. Knopf, 1984); Lynn D. Gordon, *Gender and Higher Education in the Progressive Era* (New Haven, CT: Yale University Press, 1990); Anderson, *The Education of Blacks*; Drewry and Doermann, *Stand and Prosper*; see also Joy Ann Williamson, "'Quacks, Quirks, Agitators, and Communists': Private Black Colleges and the Limits of Institutional Autonomy," *History of Education Quarterly* 44 (Winter 2004): 554–76; Cary Michael Carney, *Native American Higher Education in the United States* (New Brunswick, NJ: Transaction, 1999); Victoria-Maria MacDonald and Teresa Garcia, "Historical Perspectives on Latino Access to Higher Education, 1848–1900," in *The Majority in the Minority: Expanding the Representation of Latino/a Faculty, Administration and Students in Higher Education* (Sterling, VA: Stylus, 2003), 15–43; Philip Gleason, *Contending with Modernity: Catholic Higher Education in the Twentieth*

Century (New York: Oxford University Press, 1995); Kathleen A Mahoney, "American Catholic Colleges for Women: Historical Origins," in *Catholic Women's Colleges in America*, ed. Tracy Schier and Cynthia Russett (Baltimore: Johns Hopkins University Press, 2002); David R. Contosta, "The Philadelphia Story: Life at Immaculata, Rosemont, and Chestnut Hill," in *Catholic Women's Colleges*; Eldon L. Johnson, "Misconceptions About the Early Land-Grant Colleges," *The Journal of Higher Education* 52 (July–August, 1981): 333–51; Roger L. Geiger, "The Rise and Fall of Useful Knowledge: Higher Education for Science, Agriculture, and the Mechanic Arts, 1850–1875," *History of Higher Education Annual* 18 (1998): 47–65; Jurgen Herbst, "Nineteenth-Century Normal Schools in the United States: A Fresh Look," *History of Education* 9, no. 2 (1980): 219–27, quotation on 220; Jurgen Herbst, *And Sadly Teach: Teacher Education and Professionalization in American Culture* (Madison: University of Wisconsin Press, 1989); Christine A. Ogren, *The American State Normal School: "An Instrument of Great Good"* (New York: Palgrave Macmillan, 2005).

12. Naylor, "The Ante-Bellum College Movement," 263; Geraldine Joncich Clifford, *"Equally in View": The University of California, Its Women, and the Schools* (Berkeley: Center for Studies in Higher Education and Institute of Governmental Studies, University of California, Berkeley, 1995), 4, 6; Nancy Beadie and Kim Tolley, eds., *Chartered Schools: Two Hundred Years of Independent Academies in the United States, 1727–1925* (New York: RoutledgeFalmer, 2002); Nancy Beadie, "Internal Improvement: The Structure and Culture of Academy Expansion in New York State in the Antebellum Era, 1820–1860," in *Chartered Schools*, 89–115; Margaret A. Nash, "'A Triumph of Reason': Female Education in Academies in the New Republic," in *Chartered Schools*, 64–86; Margaret A. Nash, *Women's Education in the United States, 1780–1840* (New York: Palgrave Macmillan, 2005); David F. Labaree, *The Making of an American High School: The Credentials Market and the Central High School of Philadelphia, 1838–1939* (New Haven, CT: Yale University Press, 1988). See also William J. Reese, *The Origins of the American High School* (New Haven, CT: Yale University Press, 1995).

13. David O. Levine, *The American College and the Culture of Aspiration, 1915–1940* (Ithaca, NY: Cornell University Press, 1986), chap. 4; John L. Rury, "The Urban Catholic University in the Early Twentieth Century: DePaul, 1898–1940," *History of Higher Education Annual* 17 (1997): 5–32; Fred W. Beuttler, "Envisioning an Urban University: President David Henry and the Chicago Circle Campus of the University of Illinois, 1955–1975," *History of Higher Education Annual* 23 (2004): 107–141; Levine, *The American College*, chap. 8, quotation on 162; Steven Brint and Jerome Karabel, *The Diverted Dream: Community Colleges and the Promise of Educational Opportunity in America, 1900–1985* (New York: Oxford University Press, 1989); David F. Labaree, "The Rise of the Community College: Markets and the Limits of Educational Opportunity," chap. 8 in *How to Succeed in School Without Really Learning: The Credentials Race in American Education* (New Haven, CT: Yale University Press, 1997), 190–222, quotation on 191. See also John H. Frye, *The Vision of the Public Junior College, 1900–1940: Professional Goals and Popular Aspirations* (New York: Greenwood, 1992). For a critique of Brint and Karabel's "historical-sociological" methods, see Philo A. Hutcheson, "Reconsidering the Community College," *History of Education Quarterly* 39 (Fall 1999): 307–20.

14. Thelin, *A History of American Higher Education*, xx. Christopher Lucas' treatment of various types of higher-education institutions is similar to Thelin's; see Christopher J. Lucas, *American Higher Education: A History*, 2nd edition (New York: Palgrave Macmillan, 2006).

15. Ralph S. Brax, *The First Student Movement: Student Activism in the United States During the 1930s* (Port Washington, NY: Kennikat, 1981); Eileen Eagan, *Class, Culture and the Classroom: The Student Peace Movement of the 1930s* (Philadelphia: Temple University Press, 1981); Ronald A. Smith, *Sports and Freedom: The Rise of Big-Time College Athletics* (New York: Oxford University Press, 1988); John R. Thelin, *Games Colleges Play: Scandal and Reform in Intercollegiate Athletics* (Baltimore: Johns Hopkins University Press, 1994); Richard Nelson Current, *Phi Beta Kappa in American Life: The First Two Hundred Years* (New York: Oxford University Press, 1990).

16. Horowitz, *Campus Life*, 17.

17. Rudolph, *The American College and University*; Horowitz, *Campus Life*, chap. 2; Kim Townsend, *Manhood at Harvard: William James and Others* (Cambridge, MA: Harvard University Press, 1996); Solomon, *In the Company of Educated Women*, 95–105; Gordon, *Gender and Higher Education*. See also John Mack Faragher and Florence Howe, eds., *Women and Higher Education in American History* (New York: W. W. Norton, 1988); Elizabeth Seymour Esbach, *The Higher Education of Women in England and America, 1865–1920* (New York: Garland, 1993); Ruth Bordin, *Women at Michigan: The "Dangerous Experiment," 1870s to the Present* (Ann Arbor: University of Michigan Press, 1999).

18. Nash, *Women's Education*; Christine A. Ogren, "Where Coeds Were Coeducated: Normal Schools in Wisconsin, 1870–1920," *History of Education Quarterly* 35 (Spring 1995): 1–26, quotation on 3; Ogren, *The American State Normal School*; Kathryn Walbert, "'Endeavor to Improve Yourself': The Education of White Women in the Antebellum South," in *Chartered Schools*, 116–36, quotation on 116–17; McCandless, *The Past in the Present*; Linda Eisenmann, *Higher Education for Women in Postwar America* (Baltimore: Johns Hopkins University Press, 2006), 2; Leslie Miller-Berdal and Susan L. Poulson, eds., *Going Coed: Women's experiences in Formerly Men's Colleges and Universities, 1950–2000* (Nashville: Vanderbilt University Press, 2004). See also Laurel Thatcher Ulrich, ed., *Yards and Gates: Gender in Harvard and Radcliffe History* (New York: Palgrave Macmillan, 2004); Rosalind Rosenberg, *Changing the Subject: How the Women of Columbia Shaped the Way We Think about Sex and Politics* (New York: Columbia University Press, 2004). For further analysis of the historiography of women's higher education, see Geraldine Joncich Clifford, "'Shaking Dangerous Questions from the Crease': Gender and American Higher Education," *Feminist Issues* 3 (Fall 1983): 3–62; Patricia Palmieri, "From Republican Motherhood to Race Suicide: Arguments on the Higher Education of Women in the United States, 1820–1920," in *Educating Men and Women Together: Coeducation in a Changing World*, ed. Carol Lasser (Urbana: University of Illinois Press, 1987), 49–64; Linda Eisenmann, "Reconsidering a Classic: Assessing the History of Women's Higher Education a Dozen Years after Barbara Solomon," *Harvard Educational Review* 67 (Winter 1997): 689–717.

19. Horowitz, *Campus Life*, 51; Cally L. Waite, "The Segregation of Black Students at Oberlin College after Reconstruction," *History of Education Quarterly* 41 (Fall 2001): 344–64; Cally L. Waite, *Permission to Remain Among Us: Education for Blacks in Oberlin, Ohio, 1880–1914* (Westport, CT: Bergin and Garvey, 2003); Linda M. Perkins, "The Racial Integration of the Seven Sister Colleges," *The Journal of Blacks in Higher Education* 19 (Spring 1998): 104–108; Anderson, *The Education of Blacks*, chap. 7; Richard M. Breaux, "The New Negro Arts and Letters Movement Among Black University Students in the Midwest, 1914–1940," *Great Plains Quarterly* (Summer 2004): 147–62; Wallenstein, "Black Southerners"; Joy Ann Williamson, *Black Power on Campus: The University of Illinois, 1965–1975* (Urbana: University of

Illinois Press, 2003). See also Linda M. Perkins, "The Impact of the 'Cult of True Womanhood' on Education of Black Women," *Journal of Social Issues* 39 (1983): 17–28; Werner Sollers, Caldwell Titcomb, and Thomas A. Underwood, eds., *Blacks at Harvard: A Documentary History of African-American Experience at Harvard and Radcliffe* (New York: New York University Press, 1993); Drewry and Doermann, *Stand and Prosper*; Clayborne Carson, *In Struggle: SNCC and the Black Awakening of the 1960s* (Cambridge, MA: Harvard University Press, c. 1981, 1995); Elizabeth Higginbotham, *Too Much to Ask: Black Women in the Era of Integration* (Chapel Hill: University of North Carolina Press, 2002). On other students of color, see Carney, *Native American Higher Education*; Allan W. Austin, *From Concentration Camp to Campus: Japanese American Students and World War II* (Urbana: University of Illinois Press, 2004); Gary Y. Okihiro, *Storied Lives: Japanese American Students and World War II* (Seattle: University of Washington Press, 1999).

20. Horowitz, *Campus Life*, 95; Rury, "The Urban University," 20; Contosta, "The Philadelphia Story." Kathleen Mahoney documents the struggles of graduates of Jesuit colleges to enter post-graduate programs Harvard during the progressive era; see Kathleen A. Mahoney, *Catholic Higher Education in Protestant America: The Jesuits and Harvard in the Age of the University, 1893–1920* (Baltimore: Johns Hopkins University Press, 2003). See also Sherry Gorelick, *City College and the Jewish Poor: Education in New York, 1880–1924* (New York, NY: Schocken Books, 1982).

21. Horowitz, *Campus Life*, 68; Jana Nidiffer, "Poor Historiography: The 'Poorest' in American Higher Education," *History of Education Quarterly* 39 (Fall 1999): 321–36; Jana Nidiffer and Jeffrey P. Bouman, "'The University of the Poor': The University of Michigan's Transition From Admitting Impoverished Students to Studying Poverty, 1870–1910," *American Educational Research Journal* 41 (Spring 2004): 35–67; Herbst, "Nineteenth-Century Normal Schools," 227; Ogren, *The American State Normal School*, chaps. 3 and 5; Christine A. Ogren, "Rethinking the 'Nontraditional' Student from a Historical Perspective: State Normal Schools in the Late Nineteenth and Early Twentieth Centuries," *The Journal of Higher Education* 74 (November/December 2003): 640–64.

22. Levine, *The American College*; Beadie, "'Internal Improvement,'" 101; Nash, *Women's Education*, 99; Labaree, *The Making of an American High School*; Horowitz, *Campus Life*, 51; Leslie, *Gentlemen and Scholars*, 105, 113, 247. See also Richard Mansfield Rose, "Diverging Paths: The Emergence of Secret Societies in Antebellum Georgia Colleges," *Georgia Historical Quarterly* 70 (Spring 1986): 47–62; J. M. Opal, "The Making of the Victorian Campus: Teacher and Student at Amherst College, 1850–1880," *History of Education Quarterly* 42 (Fall 2002): 342–67. Intertwined in "the culture of aspiration" was "the culture of professionalism" outlined by Burton Bledstein; see Burton J. Bledstein, *The Culture of Professionalism* (New York: Norton, 1976).

23. Levine, *The American College*, 133; Harold Wechsler, *The Qualified Student: A History of Selective College Admissions in America* (New York: Wiley-Interscience, 1977); Karabel, *The Chosen*. See also Brint and Karabel, *The Diverted Dream*; Labaree, "The Rise of the Community College."

24. Levine, *The American College*, 169; Clyde W. Barrow, *Universities and the Capitalist State: Corporate Liberalism and the Reconstruction of American Higher Education, 1894–1928* (Madison: University of Wisconsin Press, 1990), 8–9; Suzanne Mettler, *Soldiers to Citizens: The G.I. Bill and the Making of the Greatest Generation* (New York: Oxford University Press, 2005), chaps. 3–4, quotation on 57–58; Sarah Turner and John Bound, "Closing the Gap or Widening the Divide: The Effects of the G.I. Bill and World War II on the Educational Outcomes of Black Americans," *The Journal of*

Economic History 63 (March 2003); Horowitz, *Campus Life*, 185; Daniel A. Clark, "'The Two Joes Meet–Joe College, Joe Veteran': The G.I. Bill, College Education, and Postwar American Culture," *History of Education Quarterly* 38 (Summer 1998): 165–89; David Karen, "The Politics of Class, Race, and Gender: Access to Higher Education in the United States, 1960–1986," *American Journal of Education* 99 (February 1991): 208-237, quotation on 228. See also Dongbin Kim and John L. Rury, "The Changing Profile of College Access: The Truman Commission and Enrollment Patters in the Postwar Era," *History of Education Quarterly* 47 (August 2007): 302-327. Mettler's main concern is not the types of institution the veterans attended, but rather the effect of the G.I. Bill on their sense of citizenship. Her "central finding . . . is that the G.I. Bill's education and training provisions had an overwhelmingly positive effect on male veterans' civic involvement," and "the program's effects transcended the impact of education itself." Mettler, *Soldiers to Citizens*, 9.

25. Rudolph, *The American College and University*, chap. 6; Jurgen Herbst, "The Yale Report of 1828," *International Journal of the Classical Tradition* 11 (Fall 2004): 213–31, quotation on 222; Jurgen Herbst, "American Higher Education in the Age of the College," *History of Universities* 7 (1988): 37–59; Beadie and Tolley, eds., *Chartered Schools*; Nash, *Women's Education*; Ogren, *The American State Normal School*, chaps. 1 and 3; Labaree, *The Making of an American High School*; Johnson, "Misconceptions About the Early Land-Grant Colleges"; Geiger, "The Rise and Fall of Useful Knowledge"; Anderson, *The Education of Blacks in the South*, chaps. 2 and 7; Drewry and Doermann, *Stand and Prosper*, chap. 5; see also Fred Williams, "The Second Morrill Act and Jim Crow Politics: Land-Grant Education at Arkansas AM&N College, 1890–1927," *History of Higher Education Annual* 18 (1998): 81–92. For more on science education for female academy students and the larger curricular context, see Kim Tolley, *The Science Education of American Girls: A Historical Perspective* (New York: RoutledgeFalmer, 2003).

26. Geiger, "The Era of Multi-Purpose Colleges"; Leslie, *Gentlemen and Scholars*, 80, 178; Dennis, *Lessons in Progress*, 6; Solomon, *In the Company of Educated Women*, chap. 6; Kathleen A. Mahoney and Caroline Winterer, "The Problem of the Past in the Modern University: Catholics and Classicists, 1860–1900," *History of Education Quarterly* 42 (Winter 2002): 517–43, quotation on 522; Gleason, *Contending With Modernity*; Mahoney, *Catholic Higher Education*; Julie A. Reuben, *The Making of the Modern University: Intellectual Transformation and the Marginalization of Morality* (Chicago: University of Chicago Press, 1996); Marsden, *The Soul of the American University*.

27. David F. Labaree, "Mutual Subversion: A Short History of the Liberal Arts and the Professional in American Higher Education," *History of Education Quarterly* 46 (Spring 2006): 1–15, quotation on 1; Barrow, *Universities and the Capitalist State*, 3; Brint and Karabel, *The Diverted Dream*; see also Levine, *The American College*, chap. 8; Labaree, "The Rise of the Community College." For more on vocationalism in the second half of the twentieth century, see Marvin Lazerson, "The Disappointments of Success: Higher Education After World War II," *Annals of the American Academy of Political and Social Science* 559 (September 1998): 64–76. Katherine Chaddock Reynolds' look at the roots of "Great Books" movement that began in the early twentieth century presents one small example of an effort to expand access to high status knowledge; she explains that it "represented attempts to bring liberal education to youths and adults of many social and economic classes, in essence to democratize a type of learning that had been perceived as accessible only to the intellectually elite." See Katherine Chaddock Reynolds, "A Canon of Democratic Intent: Reinterpreting

the Roots of the Great Books Movement," *History of Higher Education Annual* 22 (2002): 5–32, quotation on 7.

28. Larry Cuban, *How Scholars Trumped Teachers: Change Without Reform in University Curriculum, Teaching, and Research, 1890–1990* (New York: Teachers College Press, 1999), 9; Geiger, *Research and Relevant Knowledge*, 201, 234, 333; Jencks and Riesman, *The Academic Revolution*; Clark Kerr, "The Idea of Multiversity," in *The Uses of the University*, 5th ed. (Cambridge, MA: Harvard University Press, 2001), 1–34, quotation on 29.

29. Veysey, *The Emergence*; Marsden, *The Soul of the American University*; Reuben, *The Making of the Modern University*; Geiger, *To Advance Knowledge*, quotation on 77. On religion and scientific inquiry, see also Linda Eisenmann, "Reclaiming Religion: New Historiographical Challenges in the Relationship of Religion and American Higher Education," *History of Education Quarterly* 39 (Fall 1999): 295–306.

30. Richard M. Freeland, *Academia's Golden Age: Universities in Massachusetts, 1945–1970* (New York: Oxford University Press, 1992), 70; Geiger, *Research and Relevant Knowledge*, xix; Rebecca S. Lowen, *Creating the Cold War University: The Transformation of Stanford* (Berkeley: University of California Press, 1997), 9; Ellen W. Schrecker, *No Ivory Tower: McCarthyism and the Universities* (New York: Oxford University Press, 1986); Graham and Diamond, *The Rise of American Research Universities*, 4; Ellen Condliffe Lagemann, *The Politics of Knowledge: The Carnegie Corporation, Philanthropy, and Public Policy* (Chicago: University of Chicago Press, 1989). See also Christopher Newfield, *Ivy and Industry: Business and the Making of the American University, 1880–1980* (Durham, NC: Duke University Press, 2003); Beuttler, "Envisioning an Urban University"; Philo A. Hutcheson, *A Professional Professoriate: Unionization, Bureaucratization, and the AAUP* (Nashville: Vanderbilt University Press, 2000). Research on women in higher education has contributed monographs on faculty careers; see Margaret W. Rossiter, *Women Scientists in America: Struggles and Strategies to 1940* (Baltimore: Johns Hopkins University Press, 1982); Margaret W. Rossiter, *Women Scientists in America: Before Affirmative Action, 1940–1972* (Baltimore: Johns Hopkins University Press, 1995); Geraldine Joncich Clifford, ed., *Lone Voyagers: Academic Women in Coeducational Institutions, 1870–1937* (New York: Feminist, 1989); Patricia Ann Palmieri, *In Adamless Eden: The Community of Women Faculty at Wellesley* (New Haven: Yale University Press, 1995).

31. Jana Nidiffer and Timothy Reese Cain, "Elder Brothers of the University: Early Vice Presidents the Late Nineteenth-Century Research Universities," *History of Education Quarterly* 44 (Winter 2004): 487–523; Bashaw, *"Stalwart Women"*; Jana Nidiffer, *Pioneering Deans of Women: More Than Wise and Pious Matrons* (New York: Teachers College Press, 2000); Hutcheson, *A Professional Professoriate*; Edwin D. Duryea, *The Academic Corporation: A History of College and University Governing Boards* (New York: Falmer, 2000); Hugh Hawkins, ed., *The Emerging University and Industrial America* (Malabar, FL: Robert E. Krieger, 1985), ix–x; Newfield, *Ivy and Industry*, 82. See also Barrow, *Universities and the Capitalist State*.

32. Thelin, review of *The Emergence of the American University*, 523. See also Paul H. Mattingly, "Structures Over Time: Institutional History," in *Historical Inquiry in Education: A Research Agenda*, ed. John Hardin Best (Washington, DC: American Educational Research Association, 1983), 34–55.

33. John Aubrey Douglass, *The California Idea in American Higher Education: 1850 to the 1960 Master Plan* (Stanford, CA: Stanford University Press, 2000), 7–8, 11, 15. On the Master Plan, see also Nicholas Lemann, *The Big Test: The Secret History of the American Meritocracy* (New York: Farrar, Straus, and Giroux, 1999), book 2.

Economists Claudia Goldin and Lawrence F. Katz have turned to the history of state support for higher education between 1890 and 1940 in order to better understand current variations among states. They conclude, "The major state-level factors we identify that encouraged support to public higher education are a high level of wealth broadly distributed, the presence of business and commercial interests having large, unified, and concentrated demands for practical research, a late year of statehood, and a low early presence of private institutions. Differences across states in the relative role of the public sector in higher education that emerged during 1890–1940 persist today, despite the increased national scope of the market for higher education." Claudia Goldin and Lawrence F. Katz, "The Origins of State-Level Differences in the Public Provision of Higher Education: 1890–1940," *The American Economic Review* 88 (May 1998): 308; see also Claudia Goldin and Lawrence F. Katz, "The Shaping of Higher Education: The Formative Years in the United States, 1890–1940," *Journal of Economic Perspectives* 13 (Winter 1999): 37–62.

34. Geiger, *Research and Relevant Knowledge*; Lowen, *Creating the Cold War University*; Graham and Diamond, *The Rise of American Research Universities*, chaps. 2 and 4; Freeland, *Academia's Golden Age*, 70; Robert C. Serow, "Policy as Symbol: Title II of the 1944 G.I. Bill," *The Review of Higher Education* 27 (Summer 2004): 481–99, quotation on 493; J. M. Stephen Peeps, "A B.A. For the G.I . . . Why?" *History of Education Quarterly* 24 (Winter 1984): 513–25; Julie A. Reuben, "Merit, Mission, and Minority Students: The History of Debate Over Special Admission Programs," in *A Faithful Mirror: Reflections on the College Board and Education in America*, ed. Michael C. Johanek (New York: College Entrance Examination Board, 2001), 195–243, quotation on 197. On financial aid and affirmative action, see also Rupert Wilkinson, *Aiding Students, Buying Students: Financial Aid in America* (Nashville: Vanderbilt University Press, 2005); Lemann, *The Big Test*, book 3.

35. Geiger, *To Advance Knowledge*, 167–73; Lowen, *Creating the Cold War University*, chap. 7; Lagemann, *The Politics of Knowledge*, 6; Marybeth Gassman, "W. E. B. Du Bois and Charles S. Johnson: Differing Views on the Role of Philanthropy in Higher Education," *History of Education Quarterly* 42 (Winter 2002): 493–516, quotations on 494, 496; Eric Anderson and Alfred A. Moss, *Dangerous Donations: Northern Philanthropy and Southern Black Education, 1902–1930* (Columbia: University of Missouri Press, 1999), 11. See also Anderson, *The Education of Blacks*, chap. 7; Katrina M. Sanders, *"Intelligent and Effective Direction": The Fisk University Race Relations Institute and the Struggle for Civil Rights, 1944–1969* (New York: Peter Lang, 2005). After philanthropic support from northern industrialists began to wane in the 1930s, the presidents of several private black colleges worked together to establish the United Negro College Fund in 1944 to raise and distribute donations to their institutions; see Drewry and Doermann, *Stand and Prosper*, chap. 6; Marybeth Gassman and Edward Epstein, "Creating an Image for Black Higher Education: A Visual Examination of the United Negro College Fund's Publicity, 1944–1960," *Educational Foundations* (Spring 2004): 41–61; Marybeth Gassman, "Rhetoric vs. Reality: The Fundraising Messages of the United Negro College Fund in the Immediate Aftermath of the *Brown* Decision," *History of Education Quarterly* 44 (Spring 2004): 70–94. Andrea Walton argues that definitions of philanthropy must be broadened to include women's efforts; see Andrea Walton, ed., *Women and Philanthropy in Education* (Bloomington: Indiana University Press, 2005).

36. Johanek, ed., *A Faithful Mirror*; Harold S. Wechsler, "Eastern Standard Time: High School-College Collaboration and Admission to College, 1880–1930," in *A Faithful Mirror*, 43–79, quotation on 67; Andrea Walton, "Cultivating A Place for Selective All-Female Education in a Coeducational World: Women Educators and Professional

Voluntary Associations, 1880–1926," in *A Faithful Mirror*, 143–93, quotation on 148; Wechsler, *The Qualified Student*, xi. Linda Eisenmann also discusses the role of the AAUW in shaping women's higher education; see Eisenmann, *Higher Education for Women*, chap. 4. On the CEEB and ETS, see also Lemann, *The Big Test*, book 1; Karabel, *The Chosen*.

37. Thelin, *A History of American Higher Education*, xx.

CHAPTER 9

CURRICULUM HISTORY AND ITS REVISIONIST LEGACY

Barry M. Franklin

Educational scholars have been writing about the history of the school curriculum since the turn of the twentieth century. Yet, curriculum history has only come into its own as a distinct field of inquiry with such disciplinary trappings as a complement of identifiable practitioners, an array of investigatory methods, and a more or less shared research agenda during the last forty or so years.[1] Arriving on the scene in the late 1960s, curriculum history emerged in the midst of a movement among a group of American educational historians to reinterpret the nature and purpose of their discipline. Known as revisionism, it was an enterprise that would in various ways affect the course of development of curriculum history from its inception until the present day. The purpose of this essay is to consider the initial roots of curriculum history in the ideas of revisionism and then to look at how it has built on those origins to shape a distinct academic tradition within both educational history and curriculum studies.

The Emergence of Curriculum History

It was in the midst of the conflict over revisionism that Reese and Rury describe in chapter one that curriculum history first came into its own as a field of study. From the outset, the dividing line between these two fields of study has been somewhat unclear. It would in fact be hard to imagine how one could write about the history of education without paying attention to the curriculum and its development.[2] In a 1969 essay, Arno Bellack identified four areas of study—the development of curricular and instructional practices, the growth of curriculum as a field of work and study, the lives and careers of curriculum theorists, and the recommendations and proposals of national committees that were organized to study the curriculum—that he saw as constituting the subject matter of curriculum history. Yet in the same essay, he went on to say that curriculum history should not be a separate subject but rather a part of the general field of social and intellectual history.[3]

Although the boundaries between the two fields are unclear, it was the case that those who began writing explicitly about curriculum history in the late 1960s and early 1970s were for the most part not professional historians. They may have had some training in history or the history of education but were more likely to find themselves in the university in departments of curriculum and instruction or educational administration than in the departments of educational foundations or policy studies that were the terrain of educational historians.[4] By the 1980s, this would change, as increasing numbers of educational historians would begin to write more extensively on issues of curriculum history.[5]

Indicative of the murkiness of the boundaries between these two areas of study, it was the same conflict over revisionism that was then occurring among educational historians that would shape the development of curriculum history as a field of study. There were those curriculum historians who produced the kind of celebratory accounts that so displeased Bernard Bailyn and Lawrence Cremin. Writing in 1966, Mary Louise Seguel examined the published writings of seven individuals who she identified as the early leaders of the curriculum field, including Charles McMurry, Frank McMurry, Franklin Bobbitt, W. W. Charters, John Dewey, Harold Rugg, and Hollis Caswell. The work of these individuals, as she tells her story, established the curriculum field on an essentially progressive trajectory that over time culminated in a course of study that was attuned to the needs of a democratic society.[6]

This laudatory account of the development of the curriculum continues to exert an influence. In their 1990 history of the curriculum, Daniel and Laurel Tanner argue that the evolution of the modern American curriculum began out of the mid-nineteenth-century realization among educational reformers that the nation's schools needed a more practical course of study. What followed in their interpretation was a series of battles in which these reformers defeated a host of reactionary opponents. The result was the creation of a more accessible system of schooling and a more utilitarian school program. Each in their own way, as the Tanners see it, the founding theorists of the curriculum field contributed to an incremental and ultimately progressive transformation of the school curriculum. What first appeared on the scene as a course of study that was overly academic and remote from the lives of children became, by the beginning of the twentieth century, a curriculum that was more directly functional and related to the concerns of youth.[7] The purpose of their account, as they put it, is to "trace the great struggle for a more complete realization of the democratic potential of American society through the transformation of the school and the reformation of the curriculum."[8]

REVISIONISM AND CURRICULUM HISTORY

There were other curriculum historians who challenged this celebratory account of the development of the curriculum. These scholars have been generally sympathetic to the revisionist view of schools as instruments of social control for reproducing existing class relationships. They shared the revisionist criticism of curriculum differentiation as a means of channeling the children of the rich and poor to different courses of study, and ultimately to different and unequal life destinies.[9] Yet, they questioned the totality of the resulting regulation. They rejected the notion held by the most radical revisionist historians of education that the working classes simply accepted the direction that elites set for them through the schools. Instead, they have sought what Carl Kaestle has referred to as more "elegant" explanations of educational change that more accurately relates the roles that class conflict and consensus have played in the evolution of American public schooling.[10]

In his *Struggle for the American Curriculum*, Herbert Kliebard employs an interest group interpretive framework to examine the history of the school curriculum from the late nineteenth century to the end of the decade of the 1950s. Relying largely on national reports and the writings of influential educational leaders, the result, which is arguably the most influential and often cited account of this evolution, is an intellectual history of the conflict among four competing interest groups. There were humanists whose view of the curriculum dominated educational thought at the turn of the twentieth century. They were defenders of the classical curriculum and the doctrine of mental disciplines. Opposing them during the next half century were three other groups. There were developmentalists, such as the psychologist G. Stanley Hall, who believed that the curriculum should be organized around the interests and needs of children, and promoted the practice of child study as a means of identifying appropriate curriculum content. There was a loose collection of individuals who embraced the desire for progressive social change advocated by the sociologist Lester Frank Ward, and saw the curriculum as an instrument to redress the inequities of modern society. And there were educators who championed the doctrine of social efficiency and supported a curriculum that would socialize youth to their adult work and citizenship roles. It was only this latter group who in Kliebard's view called on the schools to assume the explicit function of social control.

There were, Kliebard argues, no victors in this struggle. Rather, what emerged in the end was what he referred to as something of a "détente" reflecting bits and pieces of the views of each of these interest groups. The resulting curriculum, as he describes it, is comprised of the traditional academic disciplines that embody our intellectual heritage. It also, however, provides for specific vocational preparation through the kind of functional courses favored by the proponents of social efficiency. It is a curriculum that includes content that serves to regulate and control children in the name of social order, while at the same time elements that free individuals from existing social constraints. As Kliebard sees it, the contemporary curriculum is comprised of subject matter that reflects the interests and desires of children as well as content that ignores those very interests in the name of supposedly greater societal needs. It is a curriculum that seems at one and the same time to promote equality and social betterment, while reinforcing existing social class divisions and the inequality that they breed. For Kliebard, then, the drive for social control is only part of the story.[11]

In *Building the American Community*, Barry Franklin offers a social control interpretation of the development of the curriculum during the first half of the twentieth century that was largely an intellectual history, but included a case study of curriculum practice in Minneapolis, Minnesota. He argues that a central concern of many American intellectuals during the early years of the twentieth century, including those who identified themselves with the development of the curriculum, was the nation's transformation from a rural, agrarian society to an urban, industrial one. Not unlike other members of the middle and upper classes of the day, these educators attributed much of the social dislocations of turn of the century America to this social transformation. Franklin goes on to argue that these early curriculum thinkers turned to the schools and their programs to restore a sense of unity and solidarity to the nation.

Where Franklin departs from the revisionist interpretation, however, is in his explanation of how these educators understood this notion of social control. There were those individuals, largely promoters of the doctrine of social efficiency, who attributed existing social disruptions to the influx into the population of an increasing number of immigrants from Eastern and Southern Europe. The schools, they argued, should become instruments of social control and create a homogeneous, like-minded culture by infusing these immigrants with the values and beliefs of the native born, Protestant middle and upper classes. There were others, John Dewey and George Herbert Mead being the best known, who also saw the school and its curriculum as an instrument of social control. Unlike their efficiency minded counterparts, they interpreted the notion of social control differently. For them, it was a process of mutual adjustment and reciprocity in which the task of the school was to create a democratic society built on the values and attitudes of all its members, native born and immigrants, workers and capitalists, Protestants and others. In his account, Franklin looks at events in Minneapolis to suggest that while the promoters of the ideas of social efficiency played a dominant role in shaping the actual school curriculum, their ability to realize their visions of social control was less clear.[12]

The impact of this revisionist brand of curriculum history has been two-fold. It spurred forward something of a radical tradition within the field itself that recognized the overtly political role that the curriculum has played as an instrument of social control. Kenneth Teitelbaum's history of socialist Sunday schools during the first two decades of the twentieth century is illustrative. These were weekend schools, according to Teitelbaum, that were established by American socialists throughout the country to offer an alternative social and political vision to what they saw as the decidedly capitalist viewpoint that their children were being taught in the regular public schools that they attended. His descriptions of teacher guides, lesson plans, readers, and other material used in these Socialist Sunday schools offers a concrete illustration of how the curriculum has been used as a political instrument. Similarly, the comparisons that he draws throughout his book with what was taught in the regular public schools that these children attended points to the equally political content of the curriculum offered in those settings.[13]

More indirectly, this revisionist orientation has provided a backdrop for the kind of critical scholarship that has come to play a major role in contemporary curriculum studies. It is not unusual for today's curriculum scholars of a critical bent to trace the origins of a host of what they see as inegalitarian school practices to the penchant for social control that revisionist oriented curriculum historians attribute to many early twentieth-century educational leaders.[14]

CASE STUDIES OF CURRICULUM HISTORY

At the outset, the focus of attention for curriculum historians was on the development of curriculum ideas and proposals.[15] This was certainly the approach that Kliebard took in his account of the development of the American curriculum. Yet, the field has not only taken its cues from the ideological debate between revisionist historians of education and their opponents. It has also been affected much like the field of educational history itself by the shift within the discipline in emphasis from intellectual to social history that accompanied the emergence of revisionism. David Labaree, one of an increasing number of educational historians who writes on curriculum history, has noted in this vein the preference of curriculum historians for examining proposals by national committees and university professors for what the schools should teach or, in his words, the "rhetorical curriculum." At least initially, it was this focus on curriculum ideas that constituted the dominant approach among those who undertook research on curriculum history. This approach, according to Labaree, has not yielded explanations that account for the development of the actual school program. What curriculum-historians need to do, he goes on to say, is to supplement their existing methods of inquiry with those that can tell us about what has in effect happened in the schools. Case studies of past instances of curriculum change and reform represent, he notes, one such methodological approach.[16]

Labaree himself has undertaken the kind of curriculum history that he advocates in his study of the development of Philadelphia's Central High School from its beginnings in 1838 until 1939. In *The Making of an American High School*, he examined a series of shifts back and forth during these years between two different curricular orientations that were directed toward contradictory goals. One course of study was terminal in nature, and was comprised of practical subjects that would prepare graduates for immediate entry into commercial and business occupations. The other program was designed to be preparatory, and was made up of academic subjects that would provide graduates with the knowledge and skills that they required for admission to college.[17] His account was one that moved the emphasis from a study of the "rhetorical curriculum" to actual school practice, as embodied in curriculum policy or what he calls the "formal curriculum," and in teachers' actual classroom practices, or what he refers to as the "curriculum-in-use."[18]

Actually, some curriculum historians had already begun to recognize the limitations that accompanied a sole focus on curriculum ideas to the exclusion of classroom and school practice. As these scholars saw it, the relationship between curriculum thought and practice was not as direct as some had thought. In most school settings there were a host of local factors that served to affect the implementation of those ideas. Political pressure, legal restraints, ideology, financial realities, and powerful personalities, to name but a few, had the ability to interfere with their reception in practice and render their impact as marginal and superficial. Like Labaree, they too suggested that curriculum historians undertake case studies of actual curriculum practice.[19]

A 1975 case study by W. Lynn McKinney and Ian Westbury that examined three curriculum reform efforts—intercultural education, science education, and vocational education—in Gary, Indiana between 1940 and 1970 is illustrative. In each instance, McKinney and Westbury noted, calls within Gary for curriculum reform and the appearance on the national scene of proposals for educational innovation had created a climate that was opportune for change. Yet, the three reform efforts that these authors described only brought about minor and short-lived modifications to the city's school program. The opposition of most of Gary's administrators and teachers to intercultural education prevented the full-scale implementation of this reform. The lack of financial resources during the 1950s and 1960s prevented Gary from implementing the kind of enriched science curriculum that was then popular on the national scene. During this same period, Gary's inadequate physical plant prevented the city from expanding the vocational program in the direction that district administrators wanted. McKinney and Westbury concluded that local conditions in Gary, including staff motivation, financial resources, and the quality of existing physical facilities, acted to limit the impact of efforts at curriculum reform.[20]

Six years later, Wayne Urban, who like Labaree is an educational historian who writes on curriculum history, examined the introduction of vocational education in the public schools of Atlanta, Georgia during the first two decades of the twentieth century. Not unlike their efficiency minded counterparts in northern cities, according to Urban, Atlanta's school reformers frequently voiced the rhetoric of vocationalism. Yet, their actual effort to vocationalize the school curriculum never achieved the success that it did in urban schools in other parts of the country. In his essay, Urban attributes this lack of change to several factors, including the support of many of Atlanta's politicians and school reformers for manual training over and above a thoroughgoing commitment to vocational education, the peculiarities of Georgia politics, and race. Taken together, these three factors severely limited the impact that vocational education had in Atlanta.[21]

In 1986, Franklin undertook a case study of curriculum change in Minneapolis to examine the efforts of school administrators in that city to introduce an integrated program organized around the interests and concerns of youth. The first such attempt occurred in 1939 with the introduction of a year long Modern Problems course in the twelfth grade to replace a required year long course in American government and a semester elective selected from sociology, economics, and commercial law. In 1945, the district embarked on a second similar initiative with the introduction of Common Learnings, a two-hour course organized around problems and concerns of youth to replace existing junior and senior high school courses in English and social studies. In this research, Franklin explored the introduction, development, and ultimate demise of both reforms, which he attributed to the effect of a number of local political and social factors that mediated the impact of proposals for reforming the schools on the actual course of study. [22]

CHALLENGING REVISIONISM

Although revisionist minded curriculum historians and their opponents interpreted the development of the curriculum in conflicting ways, they did share an essentially similar understanding about what the schools should teach. At root, both groups of these scholars were enamored of the principles of educational progressivism, particularly as those beliefs were spelled out by John Dewey, and were consequently distrustful about the practice of organizing the curriculum around the traditional disciplines of knowledge. These two groups of curriculum historians certainly did not agree about everything. They parted company, for example, on their interpretations of the work of the Committee of Ten and the Commission on the Reorganization of Secondary Education. Similarly, they disagreed about the compatibility of such school practices as curriculum differentiation with the principles of democratic education. Yet, notwithstanding these differences, both of these groups of curriculum historians looked more favorably on a curriculum that attempted to resolve current societal problems and to be responsive to the needs and interests of children than on one dedicated to using the disciplines of knowledge for passing on the cultural heritage.

Beginning in the mid-1990s however, a new viewpoint began to emerge among some of those writing curriculum history. Reflecting the pessimism of the last two decades of the twentieth century about the quality of American public schooling, these scholars held a critical view of the existing curriculum, and consequently differed with those who viewed the evolution of the curriculum in celebratory terms. At the same time, however, they rejected both the revisionist interpretive framework and the ideas of progressivism in favor of a discipline-centered course of study. One such individual has been Diane Ravitch, whose own historical account views the evolution of the twentieth century curriculum as a virtually monolithic effort on the part of professional educators to abandon a traditional discipline-centered curriculum in favor of a more child-centered one. It has been—to her way of thinking—a largely successful effort, at least heretofore, that has undermined academic standards and rigor, a commitment to genuine learning, and even a devotion to the principles of democracy and equality of opportunity.[23] Ravitch, like Kliebard, presents an intellectual history of the curriculum that focuses on similar issues and a similar cast of characters, but is decidedly more pessimistic about its long term impact on educational quality.

Such critics of revisionism and progressivism have not limited themselves to writing intellectual history. In their social history of the twentieth-century high school curriculum, David Angus and Jeffrey Mirel employed case studies of actual instances of curriculum change in two Michigan cities, Detroit and Grand Rapids, to explain the dilution of rigorous course content and the decline of academic standards for much of this period. An innovative feature of their work was the introduction of student course-taking data to bolster their argument. These records, as they saw it, pointed to a decline from the late 1920s through the early 1970s in the share of the high school curriculum devoted to academic courses, and an accompanying increase in the proportion comprising non-academic courses. They attributed this transition to the efforts of high schools, beginning during the Great Depression of the 1930s, to try to accommodate a student population that was growing in size and becoming increasingly diverse in class background and ability. School administrators of the day, Angus and Mirel claimed, began to introduce a host of more functionally oriented "personal development" courses because they believed that such subjects better matched the abilities and aspirations of their changed student population. The result over time, they argued, was the transformation of the high school from an academic institution to one whose primary function had become custodial.

Angus and Mirel were guarded about what this all meant. On the positive side, they noted that the course-taking pattern that they described seemed to have come to an end sometime during the 1970s. The ensuing two decades, they went on to say, have seen increases in graduation requirements, and a growing share of the high school curriculum being devoted to academic courses with increasing numbers of students—particularly minorities—enrolling in mathematics and science courses. On the downside, however, they also noted the creation of new courses with academic titles but with watered-down content, increases in the credit granted for completion of vocational courses, and growing patterns of curriculum differentiation. These latter changes, they argued, pose the danger of undercutting whatever efforts have been made to increase standards and enhance the academic rigor of the curriculum.[24]

THE STATE OF CURRICULUM HISTORY

As we enter the twenty-first century, much of what has characterized curriculum history research since its inception continues to define the field. In their 2005 edited volume, *Explorations in Curriculum History*, Lynn Burlbaw and Sherry Field include essays that encompass the two poles of curriculum history that I described earlier in this essay—intellectual history and social history. Much of the book is devoted to essays that examine the life and work of leading curriculum thinkers, including Alexander Inglis, Hollis Caswell, and William Bagley, and their recommendations for what schools should teach. At the same time, the volume considers what has actually occurred in and around schools—important twentieth-century curricular reform movements, the work of teachers and schools involved in various curriculum initiatives, and the interplay between major political and social events and the school curriculum.[25] Taken together, the essays in this volume suggest that as a field of inquiry, curriculum history has hardly been stagnant. Rather, new subjects—including the longstanding role of women as curriculum reformers, the influence of race on the curriculum, and the impact that international conflicts have had on the curriculum—have emerged as areas of concern for curriculum historians.[26]

The interplay between race and schooling, especially as it pertains to the education of African Americans, has proved to be an especially fruitful subject for those writing about the history of curriculum. Much of this work is broadly focused on the history of urban schools of which the curriculum is one piece.[27] For William Watkins, how-ever, the curriculum has been the central concern in his effort to frame a black conceptual lens for understanding the educational experiences of African Americans. His sources for this effort have been varied and include Marxism, radical black pedagogy, the edu-cational ideas of W. E. B. Dubois, the social-reconstructionist edu-cational vision of the 1930s, and Kliebard's interest group interpretative framework. Taken together, these differing perspec-tives lead him toward a revisionist understanding of the curriculum as it has affected black Americans from the nineteenth century onward. As he sees it, the historical role of the curriculum vis-a-vis African Americans has been an oppressive one that supports racial inequality, class divisions, and white hegemony.[28]

Craig Kridel and Vicky Newman also view curriculum history in expansive terms. As they see it, this scholarship can encompass very focused studies that look over time at proposals for what schools should teach, as well as actual instances of what schools have taught. At the same time, they believe that the field can also include more general historical examinations that consider the development of the curriculum in the context of broad social and cultural trends; the lives, careers, and recollection of important individuals; and the experiences of marginalized groups within society.[29]

A GROWING RESEARCH AGENDA

Where, then, does curriculum history appear headed as a field of study? A good portion of current research is devoted to filling and rounding out the story of the American curriculum as it has hereto-fore been developed. Craig Kridel and Robert Bullough's *Stories of the Eight Year Study* is illustrative. Their book examines the Progres-sive Education Association's twelve year effort between 1930 and 1942 to reconstruct American secondary schools, including the cur-riculum. Some of those participating in this work, Ralph Tyler and Boyd Bode for example, are well known to contemporary readers. Others, including V. T. Thayer, Wilford Aikin, and Harold Alberty, are less widely recognized. The study's successes and failings have received far less attention than the work of the 1893 Committee of Ten or the 1918 Commission on the Reorganization of Secondary

Education. Kridel and Bullough's volume, then, adds new characters and events to an already unfolding account.[30]

Similarly, curriculum historians are also extending what we currently know about the development of school subjects. The history of various academic disciplines has over the years been a popular topic for curriculum historians. The most interesting of this work has attempted in various ways to link the development of school subjects with larger social and political events.[31] A good example of this kind of connection is John Rudolph's account of discipline-centered reforms in physics and biology during the 1960s. In his book, he examines how the cold war struggle between the United States and the Soviet Union—particularly American fears about the ability of the Russians to train the engineers and scientists needed to obtain military superiority—promoted these changes. He goes on to show how such proponents of these curricular reforms as Jerome Bruner and Jerald Zacharias took what they had learned from their experience in defense-related research during World War II to the work of post-war curriculum development.[32]

Another school subject that has been of interest to curriculum historians has been special education. The first studies of this area were largely in the revisionist tradition, and sought to explore the role that special schools and classes have played as instruments of social control. In his history of special education in Boston from the 1830s through the 1920s, Robert Osgood examined the role of an array of special classes and schools in that city to provide for children who school authorities believed were difficult to teach and troublesome to manage. Most prominent among such students, Osgood goes on to say, were the increasing number of immigrant, poor, and delinquent children who were coming to populate Boston's schools. Their presence was seen by the city's educational leaders to threaten the order of the regular classroom and to disrupt the progress of its students. They further argued that these pupils needed to be segregated from other children, and provided with a more practical and less rigorous and demanding course of study. The result, Osgood concludes, was the establishment of a differentiated curriculum in these special settings that consigned their students to decidedly unequal educational and life destinies.[33]

In trying to account for the longstanding, inegalitarian role that the curriculum has played within special education, curriculum historians are beginning to look to the insights that a postmodern perspective, with its focus on language, might offer them. In a recent essay, Bernadette Baker links the discourses that surround special education that serve to identify, label, divide, and ultimately segregate disabled children to an older, nineteenth-century discourse of eugenics that played essentially the same role with populations of that day who were seen as disruptive and threatening to social order. As Baker sees it, the regulative role that special education and its curricular and pedagogical practices play are inscribed in certain ways of thinking and ways of talking about disability, schooling, and children that structure our educational institutions and practices.[34]

Baker's work is one example of an emerging body of research that employs a postmodern conceptual framework to examine the historical development of the curriculum. It is a viewpoint, Thomas Popkewitz argues, that focuses its attention on the language that we use to frame curricular proposals and practices. According to Popkewitz, embedded in the ways that we talk about curriculum and related matters are certain discursive practices that produce the systems of categories, classifications, and ordering principles that determine how we select, organize, and distribute the knowledge, skills, and dispositions that the schools teach. Curriculum history, he goes on to say, is the account of the continuities and disruptions of these discursive practices as they circulate across locations and over time.[35]

An important feature of these discursive practices, for Popkewitz, is certain regulatory or disciplinary processes that determine how children come to know who they are and how they fit into society. In his most recent work, Popkewitz focuses his attention on one important set of such cultural practices, which he calls "cosmopolitanism." These practices refer to the personal characteristics of individuals who themselves are inscribed with the Enlightenment values of reasonableness and rationality, and are capable of self-governance. These are in effect for Popkewitz the properties that set the stage for the existence of freedom, liberty, and democracy. As standards of conduct, they both discipline individuals and allow them to realize their hopes and ambitions. The question that is currently guiding Popkewitz's research is how such standards have, since the early twentieth century, become inscribed in the curriculum as the regulation of conduct is transferred from families to communities and the state.[36]

NEW DIRECTIONS FOR CURRICULUM HISTORY

Finally, what direction should curriculum history take as a field of inquiry? There are, it seems, important topics that those writing in this area have for the most part ignored. One such topic is the historical role of textbooks in shaping curriculum practice and policy. The central role that textbooks have played, and continue to play, in determining not only what is taught in schools but the pedagogy used to convey that content would suggest its importance in understanding the history of the curriculum.[37] Ironically, very little has been written about the place of textbooks in this history. William Reese's study of the origins of the American high school is an important exception. There, he notes that the curriculum of these first high schools was more likely to appear as a list of books that students were to read in preparation for their recitations than as a description of specific courses that they studied. Looking at these textbooks, then, offers us insights about the curriculum that we might not find elsewhere. Their content included the so-called modern subjects of science, English, history, and moral philosophy, and not—as is often assumed—Latin and Greek. This was a course of study, he argues, that was promoted by nineteenth-century Americans, not only as a means of mental discipline, but on utilitarian grounds as well.

The picture that Reese paints of the early high school curriculum, then, seems to challenge two widely held assumptions of many curriculum historians. First, it calls into question their belief about the supposed centrality of ancient languages in the course of study. Second, his book casts doubts about the connection that curriculum historians often make between the rise of the social efficiency movement in the early twentieth century, and the shift in emphasis to a more practically oriented course of study. As Reese sees it, an emphasis on practicality has surrounded discussions of education among Americans for a long time, and pre-dates the appearance on the scene of social efficiency thinking by almost a century.[38] Research along the lines that Reese pursues on textbooks, may serve to alter existing interpretations of the historical development of the curriculum.

Another topic that gets short shrift in current research is the curriculum after mid-century. For the most part, curriculum historians have focused their attention on educational movements and people during the first half of the twentieth century. Where they have ventured further, they typically interpret the post-World War II period as something of a replay of the preceding fifty years. Seen from this vantage point, the discipline-centered reforms of the 1960s are viewed as a recantation of the work of the Committee of Ten, the student-centered reforms of the 1970s as a the latest version of child-centered education or social reconstructionism, and today's accountability regime as the rebirth of the social efficiency movement.

What is called for, however, is a curriculum history of the last half of the twentieth century, written in its own terms. Recently, Barry Franklin and Carla Johnson have taken on this task in an essay that focuses on curriculum reform since 1950. Sub-titled "a social history of the American curriculum," they follow the increasingly popular but hardly prevailing practice of writing about what schools actually teach, as opposed to the recommendations that have been put forth for what they should teach. Such work, either with a focus on the recent past or from a social history perspective, has not been extensive. What they have found thus far, however, does suggest that some of the things that are often taken for granted about the historical development of the curriculum need to be reconsidered. At the very least, their essay reinforces the growing recognition that there is a great difference between the recommendations that national commissions and influential individuals offer concerning what should be taught and what actually takes place in the schools.

They have also found that what has occurred in the realm of curriculum change during the last fifty years is quite different than the events of the preceding half century. With the increasing involvement of the federal government and private foundations, the players participating in curriculum reform are not the same individuals who undertook this work during the early years of the twentieth century. Curriculum scholars in university schools of education—who once enjoyed a dominant role in curriculum development—have come to play something of a lesser role. They have been replaced as major players in curriculum change efforts by academics from the liberal arts as well as by non-university researchers, politicians, and ordinary citizens.

The involvement of the federal government during the early 1980s, and the resulting reform initiatives of the states, have altered the work of curriculum development in ways that were not foreseen by early twentieth-century educational reformers. Not only has this participation narrowed the scope of the curriculum to include only those content areas that are assessed by competency measures, but the role of teachers in curriculum development has become minimized. Where early twentieth-century curriculum reformers devoted much of their attention to bringing teachers into the debate over what should be taught, the involvement of today's teachers is routinely limited to selecting instructional strategies for teaching what is already dictated by the state.[39]

Curriculum history's expanding research agenda during the past four decades does point to a field that has reached something like a state of maturity. Its content is broader, and its methodologies more sophisticated. Revisionism gave birth to the study of curriculum history some forty years ago. In today's post-revisionist world, we hopefully can look forward to a curriculum history that offers a conceptually richer and deeper understanding of how Americans have, over time, approached and sought to resolve the important question that Herbert Spencer raised almost one hundred and fifty years ago, namely "what knowledge is of most worth."[40]

NOTES

1. Herbert M. Kliebard, "Constructing a History of the American Curriculum," in *Handbook of Research on Curriculum*, ed. Philip W. Jackson (New York: Macmillan, 1992), 157–84; Herbert M. Kliebard and Barry M. Franklin, "The Course of the Course of Study: History of Curriculum," in *Historical Inquiry in Education: A Research Agenda*, ed. John Hardin Best (Washington, DC: American Educational Research Association, 1983), 138–57; Barry M Franklin, "Historical Research on Curriculum," in *The International Encyclopedia of Curriculum*, ed. Arieh Lewy (Oxford: Pergamon, 1991), 63–66.

2. While there were, during the first half of the twentieth century, examinations of the history of the curriculum written by individuals whose professional identification was with curriculum in one way or another, such as Harold Rugg's essay in *The Twenty-Sixth Yearbook of the National Society for the Study of Education*, many of these first discussions of the history of curriculum appeared in larger studies of the history of public schooling. See Kliebard, 159–62; Kliebard and Franklin, fn 5, 154. Curriculum was certainly an integral part of two of the best studies of the history of twentieth century educational progressivism. See Lawrence A. Cremin, *The Transformation of the School: Progressivism in American Education, 1867–1957* (New York: Vintage Books, 1961), and Raymond E. Callahan, *Education and the Cult of Efficiency: A Study of the Social Forces that have Shaped the Administration of the Public Schools* (Chicago: University of Chicago Press, 1962).

3. Arno A. Bellack, "History of Curriculum Thought and Practice," *Review of Educational Research* 39 (June 1969): 284–91.

4. For a related discussion of the career patterns of educational historians see John Rury, "The Curious Status of the History of Education: A Parallel Perspective," *History of Education Quarterly* 46 (Winter 2006): 571–98.

5. Several of the individuals whose works are treated in this essay, including David Angus David Labaree, Jeffrey Mirel, William Reese, and Wayne Urban, are educational historians whose write on curriculum history. There are a number of other educational historians whose writings address matters that fall into the realm of curriculum history. On the education of African Americans, see James D. Anderson, *The Education of Blacks in the South, 1860–1935* (Chapel Hill: University of North Carolina Press, 1988); on gender and education, see John L. Rury, *Education and Women's Work: Female Schooling and the Division of Labor in Urban America, 1870–1930* (Albany: Sate University of New York Press, 1991), and David Tyack and Elisabeth Hansot, *Learning Together: A History of Coeducation in American Public Schools* (New Haven: Yale University Press, 1990); on vocational education, see Harvey Kantor, *Learning to Earn: School, Work, and Vocational Reform in California, 1880–1930* (Madison: University of Wisconsin Press, 1988).

6. Mary Louise Seguel, *The Curriculum Field: Its Formative Years* (New York: Teachers College Press, 1966).

7. Daniel Tanner and Laurel Tanner, *History of the School Curriculum* (New York: Macmillan, 1990), 5–6, 9–10, 133–34.

8. Ibid., xiv. For similar accounts that link the development of the curriculum to the expansion of democracy see, Peter S. Hlebowitsh, *Radical Curriculum Theory Reconsidered: A Historical Approach* (New York: Teachers College Press, 1993), 47–66, and William G. Wraga, *Democracy's High School: The Comprehensive High School and Educational Reform in the United States* (Lanham: University Press of America, 1994), 39–58.

9. For examples of the revisionist interpretation of American educational history with the emphasis on the role of schools as instruments of social control see, Michael Katz, *The Irony of Early School Reform: Educational Innovation in Mid-Nineteenth Century Massachusetts* (Boston: Beacon, 1968); Colin Greer, *The Great School Legend* (New York: Basic Books, 1972); Clarence J. Karrier, Paul Violas, and Joel Spring, *Roots of Crisis: American Education in the Twentieth Century* (Chicago: Rand McNally, 1973); Joel Spring, *Education and the Rise of the Corporate State* (Boston: Beacon, 1973); Michael Katz, *Class, Bureaucracy, and Schools: The Illusion of Educational Change in America*, exp. ed. (New York: Praeger, 1975).

10. Carl F. Kaestle, "Conflict and Consensus Revisited: Notes Toward a Reinterpretation of American Educational History: Problems and Prospects," *Harvard Educational Review* 46 (August, 1976): 4–15. One explanation for why curriculum historians were skeptical about the revisionist's interpretive framework may have been the challenges that their social control interpretation engendered among educational historians. See, for example, Diane Ravitch, *The Revisionist Revised: A Critique of the Radical Attack on the Schools* (New York: Basic Books, 1978) and David N. Plank and Paul E. Peterson, "Does Urban Reform Imply Class Conflict? The Case of Atlanta's Schools," *History of Education Quarterly* 23 (Summer 1983): 151–73. Many educational historians themselves were quick to eschew revisionism. For a discussion of this shift, see Donald Warren's introductory essay to a series of book reviews by Jeffrey Mirel, Ronald Cohen, and Michael Homel that, each in their own way, point to differences between revisionism as it was initially formulated and their brand of scholarship. See Donald Warren, "New Business in the History of Education," *Educational Studies* 18 (Spring 1987): 34–59. Another reason may have to do with the familiarity of those who challenged a celebratory interpretation of the curriculum with Edward

Krug's history of the American high school, and his effort in that book to balance notions of social control with those of social change. Krug's own background in both curriculum studies and educational history may have rendered his work particularly appealing to some curriculum historians. See Edward Krug, *The Shaping of the American High School, 1880–1920* (New York: Harper and Row, 1964).

11. Herbert M. Kliebard, *The Struggle for the American Curriculum, 1893–1958* (Boston: Routledge and Kegan Paul, 1986). Indicative of the influence of Kliebard's book and its interpretative framework is the fact that it has appeared in three separate editions— its original publication in 1986, a second and revised edition in 1993, and a third revised edition in 2004.

12. Barry M. Franklin, *Building the American Community: The School Curriculum and the Search for Social Control* (London: Falmer, 1986). The distinction that Franklin makes is between a notion of social control that sees the relationship between individuals and society as proceeding in one direction, with individuals simply inculcating predefined external societal demands and a second notion of social control that posits a two-way, reciprocal relationship between individuals and society in which individuals influence the nature of the societal demands that are placed on them. While this first type of social control typified such proponents of social efficiency as Franklin Bobbitt, W. W. Charters, and Edward L. Thorndike, it was this second from of control that was characteristic of Dewey. For Dewey's understanding of social control, see John Dewey, *Democracy and Education* (New York: Free Press, 1966). 39–40, and John Dewey, *Experience and Education* (New York: Collier Books, 1963), 51–60. For a discussion of Dewey's ideas on social control, see Robert B. Westbrook, *John Dewey and American Democracy* (Ithaca: Cornell University Press, 1991), 172, 188, and James W. Garrison, "John Dewey's Philosophy as Education," in *Reading Dewey: Interpretations for a Postmodern Generation*, ed. Larry A. Hickman (Bloomington: Indiana University Press, 1998), 77.

13. Kenneth Teitelbaum, *Schooling for "Good Rebels": Socialist Education for Children in the United States, 1900–1920* (Philadelphia: Temple University Press, 1993).

14. In his first and most important book, *Ideology and Curriculum*, Michael Apple traces the roots of the conservatism that he has argued in a number of subsequent writings dominates contemporary curriculum thought to its origins in the social efficiency movement of the early twentieth century, relying largely on the interpretations of that movement developed by revisionist oriented curriculum historians. See Michael Apple, *Ideology and Curriculum* (London: Routledge and Kegan Paul, 1979). Similarly, Thomas Popkewitz has used the work of revisionist oriented curriculum history to provide the historical context for his brand of critical curriculum scholarship. See Thomas S. Popkewitz, *A Political Sociology of Educational Reform: Power/Knowledge in Teaching, Teacher Education, and Research* (New York: Teachers College Press, 1991).

15. Barry M. Franklin, "The State of Curriculum History," *History of Education* 28, no. 4 (1999): 459–63.

16. David F. Labaree, "Politics, Markets, and the Compromised Curriculum," review of *The Struggle for the American Curriculum, 1893–1958*, by Herbert M. Kliebard, and *Building the American Community: The School Curriculum and the Search for Social Control*, by Barry M. Franklin, *Harvard Educational Review* 57 (November 1987): 493.

17. David F. Labaree, *The Making of an American High School: The Credential Market and the Central High School of Philadelphia, 1836–1939* (New Haven: Yale University Press, 1988), 134–72.

18. David F. Larabee, *Education, Markets, and the Public Good: The Selected Works of David F. Labaree* (London: Routledge, 2007), 109.

19. Kliebard and Franklin, 148–49.

20. W. Lynn McKinney and Ian Westbury, "Stability and Change: The Public Schools of Gary, Indiana, 1940–1970," in *Case Studies in Curriculum Change*, ed. William Reid and Decker Walker (London: Routledge and Kegan Paul, 1975), 1–53.

21. Wayne Urban, "Educational Reform in a New South City: Atlanta, 1890–1925," in *Education and the Rise of the New South*, ed. Ronald Goodenow and Arthur O. White (Boston: G. K. Hall and Company 1981), 114–28.

22. *Building the American Community*, 166–64. Some of Kliebard's most recent work also points to his increasing interest in case studies of actual instances of curriculum change. See the discussion of the rise of vocational education in Milwaukee, co-authored with Carol Judy Kean, in his *Schooled to Work: Vocationalism and the American Curriculum, 1876–1946* (New York: Teachers College Press, 1999), 55–118, and the examination of the emergence of the concept of curriculum in a case study of the nineteenth-century Otsego, Wisconsin Village School in his *Changing Course: American Curriculum Reform in the 20th Century* (New York: Teachers College Press, 2001), 7–23.

23. Diane Ravitch, *Left Back: A Century of Failed School Reforms* (New York: Simon & Schuster, 2000).

24. David L. Angus and Jeffrey E. Mirel, *The Failed Promise of the American High School, 1890–1995* (New York: Teachers College Press, 1999). See also, David L. Angus and Jeffrey E. Mirel, "Rhetoric and Reality: The High School Curriculum," in *Learning from The Past: What History Teaches Us about School Reform*. ed. Diane Ravitch and Maris A. Vinovskis (Baltimore: Johns Hopkins University Press, 1985), 295–328.

25. Lynn M. Burlbaw and Sherry Field, "Introduction to Explorations in Curriculum History," in *Explorations in Curriculum History*, ed. Lynn M. Burlbaw and Sherry Field (Greenwich: Information Age, 2005), 3–9.

26. Ibid. See particularly a number of essays in the volume on the work of social studies reformer Mary Kelly (161–91, 201–10); on the impact of the curriculum on black and Hispanic students (227–91); and on the role that the curriculum played in World War I and II, and in post-World War II conflicts over the influence of Communism in American life (311–28, 363–404).

27. The best and most extensive view of this relationship is provided by Jeffrey Mirel in his history of the Detroit Public Schools during the twentieth century. His book also provides a good example of how the methods of the social historian can enhance our knowledge of curriculum reform. See Jeffrey Mirel, *The Rise and Fall of an Urban School System: Detroit, 1907–81* (Ann Arbor: University of Michigan Press, 1993), chaps. 5, 6. In a recent essay, Barry Franklin looked at one of the major events that Mirel treats, the 1966 student walkout at Detroit's Northern High School, to explore how longstanding black-white conflicts over the curriculum have served to shape a sense of community among the city's African Americans. See Barry M. Franklin, "Community, Race, and Curriculum in Detroit: The Northern High School Walkout," *History of Education* 33 (March 2004): 137–56. In his recent collection of essays on the history of American urban education, John Rury includes a number of articles that explore the interplay between race and curriculum in twentieth century big city school systems. See John Rury, ed., *Urban Education in the United States: A Historical Reader* (New York: Palgrave Macmillan, 2005), parts 4 and 5.

28. William H. Watkins, "*Blacks* and the Curriculum: From Accommodation to Contestation and Beyond," in *Race and Education: The Roles of History and Society in Educating African American Students*, ed. William H. Watkins, James H. Lewis, and

Victoria Chou (Boston: Allyn and Bacon, 2001), 40–65; William H. Watkins, "A Marxian and Radical Reconstructionist Critique of American Education: Searching Out Black Voices," in *Black Protest Thought and Education*, ed. William H. Watkins (New York: Peter Lang, 2005), 107–35.

29. Craig Kridel and Vicky Newman, "A Random Harvest: A Multiplicity of Studies in American Curriculum History Research," in *International Handbook of Curriculum Research*, ed. William F. Pinar (Mawah: Lawrence Erlbaum Associates, 2003), 637–50.

30. Craig Kridel and Robert V. Bullough, *Stories of the Eight Year Study: Reexamining Secondary Education* (Albany: State University of New York Press, 2007).

31. For histories of biology, mathematics, reading, art, social studies, and other subjects that attempt to situate curriculum in a larger political and social context, see the collection of essays contained in Thomas S. Popkewitz, ed., *The Formation of School Subjects: The Struggle for Creating an American Institution* (New York: Falmer, 1987); see also George M. A. Stannic and Jeremy Kilpatrick, ed., *A History of School Mathematics*, 2 vols. (Reston: National Council of Teachers of Mathematics, 2002), and Kathleen Cruikshank, "Integrated Curriculum and the Academic Disciplines: The NCTE Correlated Curriculum of 1936," in *Curriculum and Consequence: Herbert M. Kliebard and the Promise of Schooling*, ed. Barry M. Franklin (New York: Teachers College Press, 2000), 178–96.

32. John L. Rudolph, *Scientists in the Classroom: The Cold War Reconstruction of American Science Education* (New York: Palgrave Macmillan, 2002).

33. Robert L. Osgood, *For "Children Who Vary from the Normal Type": Special Education in Boston, 1838–1939* (Washington, DC: Gallaudet University Press, 2000).

34. Bernadette Baker, "The Hunt for Disability: The New Eugenics and the Normalization of School Children," *Teachers College Record* 104 (June 2002): 663–703. For Baker's use of discursive practices for understanding the link between disability and compulsory schooling, see Bernadette Baker, "The Functional Liminality of the Not-Dead-Yet-Students, or How Public Schooling Became Compulsory: A Glancing History," *Rethinking History* 8 (March 2004): 5–49.

35. Thomas Popkewitz, "The Production of Reason and Power," *Cultural History and Education: Critical Essays on Knowledge and Schooling*, ed. Thomas S. Popkewitz, Barry M. Franklin, and Miguel Pereyra (New York: RoutledgeFalmer, 2001), 151–83.

36. Thomas S. Popkewitz, "The Reason of Reason: Cosmopolitanism and the Governing of Schooling," in *Dangerous Coagulations? The Uses of Foucault in the Study of Education*, ed. Bernadette M. Baker and Katharina E. Heyning (New York: Peter Lang, 2004), 189–223.

37. Michael W. Apple, *Teachers and Texts: A Political Economy of Class and Gender Relations in Education* (New York: Routledge and Kegan Paul, 1986), 12, 85–86.

38. William J. Reese, *The Origins of the American High School* (New Haven: Yale University Press, 1995), 103–22. For another account of the textbook in shaping the curriculum, see Jonathan Zimmerman, *Whose America? Culture Wars in the Public Schools* (Cambridge, MA: Harvard University Press, 2002), 13–31, 32–54, 55–80, passim.

39. Barry M. Franklin and Carla C. Johnson, "What the Schools Teach: A Social History of the American Curriculum since 1950," in *Handbook of Curriculum and Instruction*, ed. F. Michael Connelly (Thousand Oakes: Sage Publications), in press.

40. See Herbert Spencer, *Education: Intellectual, Moral, and Physical* (New York: D. Appleton and Company, 1897), 1–2.

CHAPTER 10

BRIDGING THE GAP BETWEEN URBAN, SUBURBAN, AND EDUCATIONAL HISTORY

Jack Dougherty

As educational history and urban history have developed in recent decades, a significant gap has opened up between them. On one side, educational historians have focused on the rise and fall of big-city school districts. On the other side, urban historians have documented how governmental housing, tax, and transportation policies fueled the postwar decline of cities and expansion of outlying suburbs. But these two fields have failed to connect with one another. In general, educational historians have not yet connected the decline of urban schools with the growth of the suburbs, and the broader political and economic shifts in the metropolitan context. Likewise, urban historians have rarely discussed what role schools played in the transformation of cities and suburbs. This chapter seeks to bridge the historiographical gap between urban, suburban, and educational history by demonstrating how these works can inform one another. It highlights major books that have served as the foundations in each field over the past few decades, as well as the rising body of new scholarship that attempts to span the distance between them.

The Rise and Fall of Urban Schools

American educational history was irrevocably altered when Bernard Bailyn delivered a wakeup call to the intellectually dormant subfield in 1960, challenging its practioners to shed their parochial views of history. His most notable target was the late Ellwood Cubberley, whose portrayal of nineteenth-century reform emphasized "great battles" for tax-supported schools and statewide control. In Cubberley's account, the forces of good (meaning the "public men of large vision") inevitably triumphed over the forces of bigotry and ignorance (such as "narrow-minded" politicians and the "old aristocratic class"). Bailyn charged that Cubberley and his contemporaries were so consumed by the reform struggles of their own generation that they wrote myopic histories, rendering the past as "simply the present writ small."[1]

Within the next decade, a new generation of educational historians answered Bailyn's call. Between 1971 and 1974, Marvin Lazerson, Michael Katz, Carl Kaestle, Stanley Schultz, and Diane Ravitch published a collection of fresh interpretations on the rise of urban school systems in nineteenth- and early twentieth-century Massachusetts and New York.[2] To be sure, this group had fierce internal divisions: the "radical revisionists" and their critics sharply disagreed on the extent to which economic determinism, social control, and human agency shaped history, and they criticized one another's interpretations and policy conclusions. But what unified this 1970s generation of scholars was their common vision that urban education systems arose as a confrontation between two cultures: the elite leaders who established institutions and the marginalized masses who they hoped would attend them.[3] Public schooling was a "battleground where the aspirations of the newcomers and the fears of the native population met and clashed," wrote Ravitch, while Kaestle described public schooling as "an institutional response to the threat of social fragmentation" due to population growth, poverty, and immigration prevalent in Northeastern cities.[4] Collectively, they replaced Cubberley's benign account with a deeper interpretation of social conflict as the driving force behind the evolution of urban schooling. In doing so, these scholars created what observers now refer to as "a 'golden age' in the historiography of city schools."[5]

The most enduring example from this golden age is David Tyack's *The One Best System: A History of American Urban Education*.[6] His book stretched far beyond New York and Boston, synthesizing case studies from across the nation into the single most comprehensive interpretation of the growth of city school systems to date. According to Tyack, elite leaders acted under the pressure of urban and industrial change to transform nineteenth-century rural village schools into twentieth-century big-city school districts, marked by greater uniformity, centralized governance, and administrative expertise. His insightful portrayal of working-class students, families, teachers, and reformers emphasized their roles as real people who took actions "inside the system," rather than passive pawns who were acted upon.

Tyack set the interpretive standard for the "post-revisionist" historical scholarship on urban education that followed, reconciling oppositional tensions from the 1970s literature. Post-revisionists generally viewed the politics of urban school reform as a "contested terrain" between administrative progressives, working-class immigrants, and racial minorities. All three groups actively supported certain reform movements, and many families sought to enroll their children in urban school systems. Although proponents of social efficiency and centralization tended to dominate debates, other forces actively proposed alternative agendas and occasionally prevailed. For example, Julia Wrigley's history of Chicago education reform identified not only business elites, but also how working-class labor leaders called for an expansion of urban schooling in ways that fit their social and economic interests. Similarly, William Reese's four-city study of Progressive era school reform emphasized the role of middle-class civic reformers who contributed to the "contested terrain" through their political struggles against administrative centralizers.[7]

While the *One Best System* stands as the classic work on the historical shift from rural to urban school systems, it scarcely mentioned another profound spatial change: suburbanization. Although Tyack brilliantly wove together several themes, he overlooked what one retrospective reviewer noted as an important one: "the territorial redistribution of the American population from cities to suburbs."[8] At the center of Tyack's narrative in 1910, the majority of the nation's population resided in rural areas and cities; only 7 percent lived in suburban areas. By the end of Tyack's epilogue in 1970, that number had reached 38 percent, meaning that a plurality of Americans lived in suburbs compared to other places. By the year 2000, the suburban population climbed to 50 percent.[9] During the same time, the urban economy also began to experience deindustrialization, beginning at the first half of the twentieth century. Manufacturers began to relocate outside of older industrial urban centers, particularly in the Northeast and Midwest, and accelerated their departure in later decades.[10]

Several historical essayists have argued that these structural changes in the metropolitan political economy have fundamentally altered the shape of public education over the twentieth century. According to Harvey Kantor, Barbara Brenzel, and Robert Lowe, as race and class divisions in the urban geography became sharper and more distinct over time, school systems could not adapt to Black and Latino demands for inclusion in the postwar era as easily as they had addressed White ethnic demands in the pre-war era.[11] Similarly, in John Rury and Jeffrey Mirel's historical overview of research traditions in the political economy of urban education, they distinguish between two major schools of thought on this twentieth-century transformation. On one hand, scholars who identify with the functionalist ecological model hold that metropolitan spatial differentiation is a natural outcome of the interaction between physical space and social inequality. On the other hand, the "new urban sociologists" insert politics into this equation, asserting that historical change is due to power conflicts between social groups whose interests are tied to specific geographical locations. While the latter model is more appealing to historians, only a handful of scholars have applied it to educational research, with only a very thin layer of supporting evidence.[12]

While many educational historians agree on the importance of twentieth-century metropolitan spatial change, it has been more difficult to illustrate this dynamic in action. For example, Ira Katznelson and Margaret Weir's *Schooling for All* boldly claims that "the possibilities of genuinely common, cross-class, cross-ethnic schooling eroded" when metropolitan areas grew during the twentieth century, because work and residence became more spatially separated, and upper and lower social classes lived further apart from one another. Previously, they assert, most children lived in the same urban school district, where struggles over governance, resources, and curriculum took place in one local political forum. When suburbanization divided the population into isolated school districts, this local forum evaporated. As a result, when increasing numbers of working- and middle-class Americans in the postwar era became able "to purchase particular kinds of public schools by purchasing specific kinds of residence areas protected by defensive zoning," Katznelson and Weir claim, "housing and schooling markets have displaced educational politics as key forums of decision making."[13] Although their argument is compelling, their case study of Chicago and San Francisco provided no direct evidence in support of this thesis about private real estate markets and public school politics. Furthermore, its nostalgic view of the early twentieth century overlooks fierce neighborhood divisions inside cities, and rural-urban conflicts over school funding in state legislatures.

Another attempt to incorporate a spatial analysis of the political economy into twentieth-century educational history is Jeffrey Mirel's study of the rise and decline of the Detroit schools. He points out that postwar suburbanization was "both a blessing and a curse" for the Motor City, as demand for cars increased, but Detroit's property tax base fell as middle-class families and factories moved out of the city during the 1950s and 1960s. Increasing numbers of black working-class families arrived in the city at the same time that its public school system had fewer resources to meet their needs. Furthermore, Mirel connects the fate of Detroit's schools to Michigan politics, where for most of the twentieth century, a rural-dominated state legislature frustrated urban attempts to secure additional educational funding. Eventually, after the U.S. Supreme Court's 1962 *Baker v. Carr* decision mandated proportional representation in the Michigan state legislature, Detroit lost seats to the booming suburbs, which tended to align into a rural-suburban block against urban interests on school finance issues.[14]

While Mirel's spatial analysis enriches our understanding of metropolitan schooling, his narrative stays focused on the rise and decline of the big-city system. The corollary rise of the suburban public schools and housing markets—and the wide socioeconomic variations among them—remains hidden in the shadowy background. From the "golden age" of urban educational history to the present, the geographical scope of our scholarship has primarily been confined to case studies of cities, and has not kept pace with the nation's suburban migration. In this respect, the current state of educational history is similar to the U.S. Supreme Court's ruling in the 1974 *Milliken v. Bradley* metropolitan school desegregation case: both stop at the city line.

URBAN DECLINE AND SUBURBAN GROWTH

Read together, the two most influential works on twentieth-century urban history tell a story of the decline of American cities and the rise of the suburbs. The first book, Arnold Hirsch's *Making the Second Ghetto*, took up the story of Chicago's housing struggles between 1940 and 1960, decades after the creation of the original African American ghettos during the Great Migration. In Hirsch's analysis, the second ghetto was formed by two factions: working-class white ethnics who violently defended their homes in racially transitional neighborhoods, and the more powerful white business elites who legally and politically manipulated Chicago's public housing and urban redevelopment agencies to relocate blacks in ways that served downtown real estate interests. "Out of the chaos emerged the second ghetto," Hirsch wrote, "an entity now distinguished by government support and sanction."[15] Two decades later, historians have remarked that Hirsch's book "changed the debate" over Northern housing discrimination by demonstrating that racial change was not caused by benign market forces, but rather by intentional public policy decisions.[16]

The second influential book, Kenneth Jackson's *Crabgrass Frontier*, revealed an urban historian's analysis of the development of the nation's suburbs, with many similarities to Hirsch. Jackson argued that postwar mass suburbanization was caused by the cultural pursuit of the "American Dream" single-family home and the racial politics of white flight, but also by governmental policies to lower housing costs (such as federally-subsidized single-family home mortgages and interstate highway construction).[17] Once again, suburbia did not occur simply due to "natural" market forces, but was the intended result of public policy decisions. Jackson's book inspired a new generation of urban historians to make sense of suburbs, sparking the creation of what proponents have labeled "the new suburban history" to adapt his interpretation to a wider variety of settings, including African American and working-class suburbs.[18] Scholars like Amanda Seligman have commented on the intimate connections between Hirsch's and Jackson's accounts of urban decline and suburban growth, noting that governmental actions financed African American containment in one sector and white expansion in another. She and others have called for redefining the fields of urban and suburban history into a consolidated "metropolitan history," to more clearly signify the intellectual linkage between them.[19]

Perhaps the most widely recognized exemplar in this new field is Robert Self's *American Babylon*, which creatively ties the decline of Oakland, California to the growth of East Bay suburbs during the postwar era. Self argues that the black power struggle and the Proposition 13 tax revolt are actually two halves of the same urban-suburban story. While the Black Panthers demanded governmental policies to benefit the impoverished residents of Oakland, conservative suburbanites responded by passing property tax caps that sharply curtailed their fiscal responsibilities to state and local governments. By explicitly linking urban and suburban narratives on a wide range of topics, including housing, labor, public services, and civil rights, Self brings these two fields much closer together.[20]

But in Self's otherwise comprehensive account of metropolitan history, what is especially striking is the virtual absence of any discussion of schools. According to one reviewer, *American Babylon* mentions public education only once, when quoting a white East Oakland resident who refused to send her children to a school with "too many colored" students.[21] Looking back on the classics in urban and suburban history, perhaps this absence should not surprise us. Hirsch's *Making the Second Ghetto* barely mentions schooling at all, and in Jackson's *Crabgrass Frontier*, it appears in only a few paragraphs, isolated from the central narrative of the book.[22] The gap between these different fields of history—on both sides—is remarkable. Whereas educational historians tend to stop at the city line, urban and suburban historians appear to have stopped at the schoolhouse door.

EXPLAINING THE GAP BETWEEN CITIES, SUBURBS, AND SCHOOLS

Why does this divide exist between educational history and urban-suburban history? One reason may be because the conventional interpretations from each field do not neatly fit alongside one another. For example, in Kenneth Jackson's thesis on why mass suburbanization happened, he claims that there were "two necessary conditions . . . the suburban ideal and population growth —and two fundamental causes—racial prejudice and cheap housing." Expanding on the racial prejudice theme, Jackson briefly discusses the role of schooling. In the wake of the 1954 *Brown* school desegregation case, he claims that "millions of families moved out of the city 'for the kids' and especially for the educational and social superiority of smaller and more homogenous suburban school systems."[23]

Indeed, white flight to suburban schools did occur, but not according to the compressed chronology that Jackson offers here. During the late 1940s and 1950s, schooling had not yet become a primary motivation for suburban migration. Based on David Tyack's *The One Best System,* we know that most urban districts were still recognized as the nation's prized exemplars of public education in this era, with physical facilities and services that typically surpassed what less-densely populated areas could offer. Furthermore, according to Herbert Gans' sociological study of the Levittown, New Jersey suburban development, which opened in a sparsely settled agricultural area near Philadelphia in 1958, less than 1 percent of residents cited schooling as a reason for leaving their previous residence or selecting this new community. Yet these families cared a great deal about the quality of public education. Gans devoted an entire chapter to the intense conflicts he observed between Levittown's rural school superintendent (who provided a traditional, basic education) versus the newly arrived middle-class suburbanites (who demanded a more challenging and expensive curriculum to prepare their children for prestigious colleges and universities).[24]

Schooling does not fit neatly into Jackson's suburbanization thesis because its role reverses during the late twentieth century. Although typical suburban schools did not attract families during the 1940s and 1950s, they eventually became an extremely strong magnet in the 1970s and 1980s. During this later period, more families left cities expressly to enroll their children in suburban schools, despite the fact that suburban housing costs were no longer as affordable as they had been a few decades earlier. Suburban schools flipped—from a negligible factor to an extremely influential motivator—halfway through the great white migration of the late twentieth century. In addition, school finance battles became more contentious in many state legislatures and courts as dollar costs for increasingly competitive schools rose sharply, and districts were torn between offering what newer residents demanded versus what older residents had settled for in their day. We need richer histories of cities, suburbs, and schooling to fully understand when, where, and how these transformations—which could alter interpretations in both the educational and urban-suburban literature—occurred.

Bridging the gap also would help to reconcile some of the differences between case studies that fail to connect with one another. For example, consider two different prize-winning interpretations of postwar Detroit: one by an educational historian (Jeffrey Mirel's *The Rise and Fall of an Urban School System*) and the other by an urban historian (Thomas Sugrue's *Origins of the Urban Crisis*).[25] On one hand, Mirel argues that a liberal-labor-black political coalition rose up to bolster Detroit schools in the 1950s and early 1960s, but its collapse in the later 1960s signaled the rapid decline of the district. On the other hand, Sugrue focuses on intense racial conflicts in housing and employment prior to the 1960s, thereby casting doubt on whether a liberal-labor-black coalition actually existed as described in Mirel's book. How do we deal with these seemingly incompatible interpretations? Is it possible that a cross-racial coalition was formed on some civil rights issues (like schooling) but not others (like housing and jobs)? This question remains unanswered. Neither of these books, published just three years apart, cites the other author's work, nor previous journal articles by him.[26] Furthermore, not a single historical journal has published a review that compares both Mirel's and Sugrue's interpretations. With enormous gaps like this between the literatures of educational history and urban-suburban history, both of our fields suffer.

EXEMPLARS FOR BRIDGING THE DIVIDE

Within the past few years, a growing body of scholarship has begun to bridge the gap between cities, suburbs, and schools. Some works have been authored by educational historians, some by urban or suburban historians, and some by social scientists doing thoughtful historical research. The examples offered here are intended to recognize interesting work by a rising generation of scholars, and to inspire others about the range of possibilities.

One strand of new scholarship looks more closely at the connections between private real estate and public schools. Kevin Fox Gotham, in his insightful study of Kansas City, Missouri, explains why traditional factors (such as urban renewal, interstate highways, and migration patterns) fail to explain the shape of racial change from the 1950s to the 1970s. Instead, he argues, one needs to understand the political economy of public education and housing. Gotham identifies how school administrators, real estate agents, and community activists struggled over "the unwritten law of the Troost line," a racial boundary separating schools and neighborhoods on either side of a major avenue dividing the city.[27] In a related study, Amanda Seligman demonstrates the role that residential "block-busting" played in reshaping Chicago by altering racial housing patterns and neighborhood schools.[28] Real estate also involves the physical space upon which schools are located. Michael Clapper's research analyzes the political decisions and cultural meanings behind the site selections and architecture of public and parochial schools in Philadelphia and its suburbs between 1945 and 1975. Using archives, oral histories, and computer mapping, it traces how the construction of public school buildings solidified inequalities across the metropolitan region.[29]

A second strand of literature bridges the gap by emphasizing urban and suburban residents' cultural ties to schools. For example, Gerald Gamm's study of Boston neighborhoods investigates why the Jewish exodus to the suburbs occurred earlier and faster than in Catholic neighborhoods, reminding us that the generic phrase "white flight" does not capture important variations. His detailed community study of racial succession explains how Catholics identified more closely with neighborhood institutions—such as parochial schools and churches—which created a stronger sense of neighborhood stability amid racial transition.[30] Baxandall and Ewen's rich portrait of Long Island, New York traces suburbanization from its all-white origins to its present-day racial diversity. Their book also connects heated political battles over public-private housing and racial segregation with changes in the fabric of suburban community life, particularly women whose life stories were framed in part by the newly constructed suburban schools their children attended.[31] On a related theme, Claudia Keenan's study of two different suburban communities in metropolitan New York City examines the cultural lives of women and men through their parent-teacher associations, and their role in defining suburban lifestyles that equated "good schools" with "the good life."[32]

A third strand of new scholarship moves outside of the stereotypical Northeast and Midwest to explore the range of city-suburban dynamics involving schools in the West and the South. In her study of Compton, California, historian Emily Straus analyzes the external and internal factors that transformed the public schools of this comfortable Los Angeles suburb of the 1950s into an "urban crisis" by the 1980s. One factor, she argues, was the Compton's residents "held onto the ideal of suburbia" as fiscal resources tightened around their community, eventually leading to a dual decline in school quality and property values.[33] Related themes also appear in portions of Becky Nicolaides' history of the working-class Los Angeles suburb of South Gate, which shifted from a Depression-era democratic stronghold to a civil rights-era base of white conservatism. Nicolaides argues that South Gate's residents' primary concerns about homeownership and taxes were expressed most dramatically in the politics of race and education, both in the 1930s and the 1960s, with different results in each period.[34]

In many southern states, school district boundaries were drawn at the county level, meaning that city-suburban tensions occurred within one large metropolitan area. In *Boom for Whom?* political scientist Stephen Smith explains how the white business elite supported school desegregation to promote economic development, more so than educational equity, in Charlotte-Mecklenburg, North Carolina.[35] Historian Ashley Erickson is writing a study of Nashville-Davidson County, Tennessee, which pays close attention to how real estate interests influenced desegregation planning, and its effects on school site locations and student curricula.[36] Kevin Kruse's book on "white flight" in metropolitan Atlanta argues that school desegregation reshaped the urban-suburban Sun Belt as much as deindustrialization affected the Rust Belt.[37] Finally, Matthew Lassiter's regional study of the metropolitan South explores the rise of a "color-blind" ideology as middle-class white suburbanites reacted against racial desegregation by defending what they viewed as their natural entitlement to neighborhood schools.[38]

A fourth strand of new scholarship seeks to draw connections between cities, suburbs, and schools by focusing on historical changes in two closely related markets: education and housing. My own work on metropolitan Hartford, Connecticut investigates how middle-class Americans increasingly began to "shop around" for the best schools in postwar suburbia, thereby transforming public education into a commodity to be bought and sold through the private real estate market. I argue that "shopping for schools" became more widespread as accumulating educational credentials for one's children became a more reliable route toward socioeconomic mobility in the human capital labor market of the mid-twentieth century. Although governmental policy remains a key player, this study also presents evidence on the school-home market from both the sellers' and buyers' perspectives. During the postwar era, realtors increasingly featured selected suburban schools in advertisements and promotional materials, and homebuyers reported school quality as a greater motivation in purchasing decisions. Furthermore, local town officials became more heavily involved in both cooperation and conflicts with realtors, residents, and "outsiders" amid the changing relationship between schools and housing.[39] This scholarship draws on prior work by David Labaree, who traced the origins of an academic credentials market for elite public high schools back to the late nineteenth century, and also work by Lizabeth Cohen and others who have richly documented the mass consumer culture expansion of the postwar era.[40] It also seeks richer sources of evidence to test ideas originally raised in Katznelson and Weir's *Schooling for All*, about changes in metropolitan space and the politics of education.

CONCLUSION

Historical writing reflects a great deal about changes occurring during the context in which it was authored. In the 1970s, during the "golden age" of educational history, scholars sought to understand the role of education in nineteenth-century cities, and perhaps any insights they might offer regarding the urban school protests over their own generation. Similarly, in the 1980s, leading urban and suburban historians closely examined how governmental actions of the postwar era created the unequal social geography that had become more apparent in their own period. Today, scholars from both fields are beginning to make sense of how cities, suburbs, and schooling came together and influenced one another during the twentieth century, producing the results that are so evident to our own eyes.

Collectively, these strands can improve the quality of our scholarship by bringing educational history and urban-suburban history closer together, so that ideas and evidence from both fields may interact with one another. But this is not solely an academic matter. Better histories of cities, suburbs, and schooling also have the potential to contribute to broader public policy discussions on this controversial and complex topic. Over four decades ago, James Bryant Conant published *Slums and Suburbs*, and broad audiences continue to read popular accounts on the same theme by Jonathan Kozol and other activists and journalists.[41] Furthermore, school finance debates that pit the interests of different types of school districts against one another continue to grow across most of the nation's fifty states. The topic of cities, suburbs, and schooling clearly matters to the American public. Although history will not solve our contemporary policy dilemmas, it can give us a clearer sense of how we arrived at this point, and perhaps some ideas about our next steps.

NOTES

1. Ellwood P. Cubberley, *Public Education in the United States: A Study and Interpretation of American Educational History; Revised and Enlarged Edition* (Boston: Houghton Mifflin, 1919/1947), 164–65; Bernard Bailyn, *Education in the Forming of American Society; Needs and Opportunities for Study* (Chapel Hill: University of North Carolina Press, 1960), 9.

2. Marvin Lazerson, *Origins of the Urban School: Public Education in Massachusetts, 1870–1915* (Cambridge, MA: Harvard University Press, 1971); Michael Katz, *Class, Bureaucracy, and Schools: The Illusion of Educational Change in America*, rev. ed. (1971; repr., New York: Praeger, 1975); Carl F. Kaestle, *The Evolution of an Urban School System: New York City, 1750–1850* (Cambridge, MA: Harvard University Press, 1973); Stanley Schultz, *The Culture Factory: Boston Public Schools, 1789–1860* (New York: Oxford, 1973); Diane Ravitch, *The Great School Wars, New York City, 1805–1973: A History of the Public Schools as Battlefield of Social Change* (New York: Basic Books, 1974).

3. Kaestle, *Evolution of an Urban School System*, 188.

4. Ravitch, *The Great School Wars*, xiii; Kaestle, *Evolution of an Urban School System*, viii.

5. John L. Rury, ed., *Urban Education in the United States: A Historical Reader* (New York: Palgrave Macmillan, 2005), 2; Ruben Donato and Marvin Lazerson, "New Directions in American Educational History: Problems and Prospects," *Educational Researcher* 29 (November 2000): 5–6.

6. David Tyack, *The One Best System: A History of American Urban Education* (Cambridge, MA: Harvard University Press, 1974).

7. Julia Wrigley, *Class Politics and Public Schools: Chicago 1900–1950* (New Brunswick, NJ: Rutgers University Press, 1982); William J. Reese, *Power and the Promise of School Reform: Grassroots Movements During the Progressive Era* (Boston: Routledge and Kegan Paul, 1986).

8. Harvey Kantor, "In Retrospect: David Tyack's *The One Best System*," *Reviews in American History* 29 (2001): 326. Tyack's book examined post–World War II demographic and economic changes in cities (*One Best System*, 276–78), but did not discuss how suburbanization altered the politics of education.

9. Frank Hobbs and Nicole Stoops, *Demographic Trends in the 20th Century: Census 2000 Special Reports, Series Censr-4* (Washington, DC: U.S. GPO, 2002), 33.

258   JACK DOUGHERTY

10. Jean Anyon, *Ghetto Schooling: A Political Economy of Urban Educational Reform* (New York: Teachers College Press, 1997), chap. 4.

11. Harvey Kantor and Barbara Brenzel, "Urban Education and the 'Truly Disadvantaged': The Historical Roots of the Contemporary Crisis, 1945–1900," *Teachers College Record* 94 (Summer 1992): 278–314; Robert Lowe and Harvey Kantor, "Considerations on Writing the History of Educational Reform in the 1960s," *Educational Theory* 39 (1989): 1–9.

12. John L. Rury and Jeffrey E. Mirel, "The Political Economy of Urban Education," in *Review of Research in Education, Volume 22*, ed. Michael W. Apple (Washington, DC: AERA, 1997) 49–110.

13. Ira Katznelson and Margaret Weir, *Schooling for All: Class, Race, and the Decline of the Democratic Ideal* (Berkeley: University of California Press, 1985/1988), 27, 215.

14. Jeffrey Mirel, *The Rise and Fall of an Urban School System: Detroit, 1907–1981* (Ann Arbor: University of Michigan Press, 1993), 218–21, 244–50, 294–98.

15. Arnold R. Hirsch, *Making the Second Ghetto: Race and Housing in Chicago, 1940–1960* (Cambridge, MA: Cambridge University Press, 1983), xiii.

16. Timothy J. Gilfoyle, "Introduction: Urban History, Arnold Hirsch, and the Second Ghetto Thesis," *Journal of Urban History* 29 (March 2003): 233.

17. Kenneth T. Jackson, *Crabgrass Frontier: The Suburbanization of the United States* (New York: Oxford University Press, 1985).

18. Kevin M. Kruse and Thomas J. Sugrue, eds. *The New Suburban History* (Chicago: University of Chicago Press, 2006); Becky M. Nicolaides and Andrew Wiese, eds. *The Suburb Reader* (New York: Routledge, 2006).

19. Amanda I. Seligman, "What is the Second Ghetto?," *Journal of Urban History* 29 (March 2003): 274; Amanda I. Seligman, "The New Suburban History [Review Essay]," *Journal of Planning History* 3 (November 2004): 312–23.

20. Robert O. Self, *American Babylon: Race, Power, and the Struggle for the Postwar City in California* (Princeton: Princeton University Press, 2003).

21. Michael Flamm, "Destructive Winds", *Reviews in American History* 32, no. 4 (2004): 552–57; the reference to public education in *American Babylon* is on page 166.

22. The topic of schooling appears very briefly in Hirsch, *Making the Second Ghetto*, 65, 233. In Jackson, *Crabgrass Frontier*, only three paragraphs link schooling and suburban growth, 147, 289–90.

23. Jackson, *Crabgrass Frontier*, 289–90.

24. Herbert J. Gans, *The Levittowners: Ways of Life and Politics in a New Suburban Community* (New York: Pantheon, 1967), 31–41, 86–103.

25. Mirel, *The Rise and Fall of an Urban School System*; Thomas F. Sugrue, *The Origins of the Urban Crisis: Race and Inequality in Postwar Detroit* (Princeton: Princeton University Press, 1996).

26. Mirel's book was published first in 1993. Sugrue's book (1996) does not cite Mirel, nor does Mirel's second edition (1999) cite Sugrue.

27. Kevin Fox Gotham, *Race, Real Estate, and Uneven Development: The Kansas City Experience, 1900–2000* (Albany, NY: SUNY, 2002), chap. 5.

28. Amanda I. Seligman, *Block By Block: Neighborhoods and Public Policy on Chicago's West Side* (Chicago: University of Chicago Press, 2005), chap. 5.

29. Michael Clapper, "Separate and Unequal: The Constructed World of Philadelphia Area Schools After 1945" (PhD diss., University of Pennsylvania, expected 2007).

30. Gerald Gamm, *Urban Exodus: Why the Jews Left Boston and the Catholics Stayed* (Cambridge, MA: Harvard University Press, 1999).

31. Rosalyn Baxandall and Elizabeth Ewen, *Picture Windows: How the Suburbs Happened* (New York: Basic Books, 2000).
32. Claudia Keenan, "P.T.A. Business: A Cultural History of How Suburban Women Supported the Public Schools, 1920–1960" (PhD diss., New York University, 2002).
33. Emily Straus, "The Making of the American School Crisis: Compton, California and the Death of the Suburban Dream." (PhD diss., Brandeis University, 2006), 3.
34. Becky M. Nicolaides, *My Blue Heaven: Life and Politics in the Working-Class Suburbs of Los Angeles, 1920–1965* (Chicago: University of Chicago Press, 2002), 156–68, 286–94.
35. Stephen Samuel Smith, *Boom for Whom? Education, Desegregation, and Development in Charlotte* (Albany, NY: SUNY, 2004). See also Gregory S. Jacobs, *Getting Around Brown: Desegregation, Development, and the Columbus Public Schools* (Columbus: Ohio State University Press, 1998).
36. Ansley Erickson, "Schooling the Metropolis: The Growth Ideal and Educational Inequity, Nashville, TN, 1950–2000" (PhD diss., Columbia University, in progress).
37. Kevin Michael Kruse, *White Flight: Atlanta and the Making of Modern Conservatism* (Princeton, NJ: Princeton University Press, 2005).
38. Matthew D. Lassiter, *The Silent Majority: Suburban Politics in the Sunbelt South* (Princeton, NJ: Princeton University Press, 2006).
39. Jack Dougherty, "The Rise of 'Shopping for Schools' in Suburbia." Paper to be presented at the History of Education Society meeting, October 2007. See the "Cities, Suburbs, and Schools" research project at Trinity College website: http://www.trincoll.edu/depts/educ/css.
40. David F. Labaree, *The Making of an American High School: The Credentials Market and the Central High School of Philadelphia, 1838–1939* (New Haven, CT: Yale University Press, 1988); Lizabeth Cohen, *A Consumers' Republic: The Politics of Mass Consumption in Postwar America* (New York: Knopf, 2003). See also Kimberley Tolley, "Learning in a Consumers' Republic," *History of Education Quarterly* 46, no. 2 (2006): 274–88.
41. James Bryant Conant, *Slums and Suburbs: A Commentary on Schools in Metropolitan Areas* (New York: McGraw Hill, 1961); Jonathan Kozol, *Savage Inequalities: Children in America's Schools* (New York: Crown, 1991).

CHAPTER 11

THE FEDERAL ROLE IN
AMERICAN EDUCATION:
A HISTORIOGRAPHICAL ESSAY

Adam R. Nelson

Twenty-five years ago in the *Harvard Educational Review*, Carl Kaestle and Marshall S. Smith published their article, "The Federal Role in Elementary and Secondary Education, 1940–1980." Noting the gradual process of state-level centralization that had occurred in American public education over the course of the nineteenth century, they suggested that a remarkably similar process of centralization had occurred at the federal level since the middle of the twentieth century. "One of the tasks of historical analysis," they observed, "is to determine whether the increase in federal involvement represents a continuation of a long-range process of centralization in education." Was this increased centralization at the federal level historically inevitable, they asked, or was it perhaps "reversible"?[1]

Their short answer was that, despite attempts in the 1980s to shift the momentum and return power to states, centralization at the federal level was likely to continue. "The vague anxieties that many educators and laypeople had about federal involvement in the 1940s and 1950s have been replaced by hot debate about the efficacy

262 ADAM R. NELSON

and desirability of the federal government's efforts to achieve equity in local school systems," they argued.[2] By the 1980s, a prominent—and growing—federal role had become a central feature of the American educational landscape, and subsequent developments have only confirmed this trend.

In the decades since Kaestle and Smith published their initial article, historians of the federal role have significantly broadened and deepened (but have not fundamentally revised) this interpretation. This chapter surveys the historical literature on "the federal role" and comments on its strengths and weaknesses. Just as the federal role in education has evolved over time, so too has the historiography. What remains to be seen is whether the steady accumulation of historical analyses that have emerged in recent decades will yield any consensus about the overall significance of a growing federal role in American public schools.

Very few historians of the federal role attempt comprehensive interpretations of the subject. As with most fields of historical study, scholars focus on discrete subtopics. For example, they might investigate the development of federal aid for the economically disadvantaged, or federal guidelines for teaching the disabled, or federal interventions on behalf of English language learners. Rarely do historians examine all of these groups—and their related federal programs—simultaneously. Instead, historians choose to write about specific federal programs and specific groups of students. Histories exploring the interactions among multiple programs or multiple constituencies are rare.[3]

Similarly, few historians of the federal role in education deal with more than one branch of the federal government at a time. Some write legislative histories that focus on the inner workings of Congress. Others write "implementation" studies that focus heavily on the regulatory agencies of the executive branch (many of which, from the Department of Education to the Department of Justice, oversee school-related programs). A few stress the interplay between the legislative and executive branches, and a great many emphasize the influence of the federal courts, but seldom do historians bring all three branches of the federal government together in a single, nuanced, multi-layered analysis.

Even when historians attempt to deal with all three branches of government, they often find it difficult to draw connections between the "federal" narrative and the diverse local settings in which federal policies operate. This difficulty is understandable given the huge array of local factors that affect policy implementation on the ground, but failing to situate federal policies in local context may prevent historians from conveying the actual impact of the federal role in ordinary citizens' lives. Some historians therefore use local case studies to document the complex give-and-take that occurs between local, state, and federal actors over time—but, of course, case studies involve drawbacks of their own.

By and large, histories of the federal role in American education fall roughly into two periods: *before* and *after* World War II. Studies addressing the period before World War II examine such topics as the Northwest Ordinances of 1785 and 1787 (which gave federal land to support new schools in frontier territories), the Morrill Act of 1862 (which gave land to support "agricultural and mechanical" colleges), and the Freedmen's Bureau (which supported the education of former slaves during Reconstruction). Despite various constitutional objections to federal aid, these early grants, though relatively small, played a key role in building the nation's educational infrastructure in the nineteenth century.[4]

Of course, traditionally, education in the United States has been considered a local and state responsibility, and fears of "federal control" have a long history. Gordon Lee's book *The Struggle for Federal Aid: The First Phase* documents how debates over federal control derailed both the Hoar Bill for grants-in-aid in the 1870s and the Blair Bill in the 1890s. Similar fears arose in the early twentieth century in legislative debates over the Smith-Lever Act of 1914 (which offered aid to agricultural extension programs) and the Smith-Hughes Act of 1917 (which funded vocational training classes). Herbert Kliebard examines these laws in *Schooled to Work: Vocationalism and the American Curriculum, 1876–1946*.[5]

Other histories of the federal control debate include Gilbert E. Smith, *The Limits of Reform: Politics and Federal Aid to Education, 1937–1950*, as well as Richard Reiman, *The New Deal and American Youth: Ideas and Ideals in a Depression Decade*. Both volumes note the ways in which federal officials in the 1930s deliberately by-passed state and local officials to set up education-related services under the Civilian Conservation Corps (CCC) and the National Youth Administration (NYA). As Smith explains, the CCC and NYA—run wholly under federal control—justified this approach by citing an urgent need for workforce development and poverty relief in a period of national emergency.[6]

The idea of federal aid to meet "emergency" needs in U.S. education resurfaced, of course, after World War II. A widely cited work on federal-aid debates in the 1950s is Barbara Barksdale Clowse's detailed legislative history, *Brainpower for the Cold War: The* Sputnik *Crisis and National Defense Education Act of 1958.* Tracing the day-to-day political negotiations that occurred among key figures in Congress and the White House before and after the launching of *Sputnik*, Clowse notes the strategies Eisenhower officials used to sustain a sense of "crisis" in education and thus to win support for large-scale federal aid to schools.[7]

Repeated attempts in the 1950s and early 1960s to boost federal aid to schools led to new works on federal education politics—including Robert Bendiner, *Obstacle Course on Capitol Hill*; Virgil Blum, *Freedom in Education: Federal Aid for All Children*, which dealt extensively with the debate over federal aid to private and parochial schools; and Sidney Tiedt, *The Role of the Federal Government in Education.* Yet, it was the passage of the landmark Elementary and Secondary Education Act of 1965 (the ESEA) that ultimately redirected historical scholarship on the federal role in the nation's schools.[8]

The unprecedented size and scope of the ESEA led many to ask how the federal government would be able to manage such a sprawling program. Contemporary histories such as Stephen Bailey and Edith Mosher, *ESEA: The Office of Education Administers a Law*, and Eugene Eidenberg and Roy Morey, *An Act of Congress: The Legislative Process and the Making of Education Policy*, questioned the relationship between the ESEA's ambitious legislative goals and its promised educational results. Increasingly, old debates over federal control gave way to new debates over the *effectiveness* of large-scale federal aid to schools.[9]

The standard legislative history of the ESEA is Julie Roy Jeffrey's *Education for Children of the Poor: A Study of the Origins and Implementation of the Elementary and Secondary Education Act of 1965.* Jeffrey's narrative positions confident Great Society policymakers against critics who doubted that federal grants to "compensatory" programs for low-income pupils could achieve the Johnson administration's stated goal of eradicating poverty. Although early assessments of the ESEA proved disappointing—both in terms of reducing poverty and in terms of raising student achievement—Jeffrey observed that ESEA programs nonetheless won broad political support.[10]

Why did the ESEA become so popular so quickly? Like others who have studied the political longevity of Great Society programs, Jeffrey highlighted the categorical (i.e., carefully targeted) nature of ESEA programs and their capacity to attract well-organized constituencies that ensured a steady flow of aid to specific groups. Indeed, the late 1960s and early 1970s saw a proliferation of so-called "iron triangles," or tight policy networks linking interest groups, regulatory offices, and congressional committee staffers around categorical programs serving highly mobilized and politically savvy constituencies.

The pros and cons of categorical grants have had a particularly complex history in the case of Head Start, the federally funded preschool program. Early proponents of Head Start endeavored to solve the problem of "program capture" by linking Head Start (based outside the federal Office of Education) with various community-based services. Yet, as Maris Vinovskis has shown in *The Birth of Head Start: Preschool Education Policies in the Kennedy and Johnson Years*, the diffuse goals of the program left supporters without a clear measure of its effectiveness. Eventually, the political advantages of using strictly academic criteria to assess Head Start displaced more holistic assessments.[11]

Attempts to "assess" categorical grant programs led to a robust literature among policy analysts in the early 1970s. For example, when the Nixon administration set out to measure the academic outcomes of federal-aid programs, Jerome T. Murphy published "Title I of ESEA: The Politics of Implementing Federal Educational Reform," *Harvard Educational Review*. This article, which echoed the findings of Ruby Martin and Phyllis McClure in their critique, *Title I of ESEA: Is It Helping Poor Children?*, revealed that federal officials in the 1960s had been woefully underprepared to monitor the results of federal programs, resulting in rampant misuses of Title I funds.[12]

Evaluating the effectiveness of categorical programs became a key preoccupation of education scholars in the 1970s and launched a new debate over "accountability" in the public schools. Books such as Edward Wynne's *Politics of School Accountability: Public Information About Public Schools* set the stage for new debates over the pros and cons of "market competition" among schools, as did Jerome Murphy and David Cohen's case study, "Accountability in Education: The Michigan Experience," *The Public Interest*. This debate over accountability soon reframed historians' approach to the federal role: increasingly, historians wanted to know whether federal programs had "worked."[13]

About this time, Jeffrey Pressman and Aaron Wildavsky published their widely reviewed book, *Implementation: How Great Expectations in Washington are Dashed in Oakland; or Why It's Amazing That Federal Programs Work At All*, and Milbrey McLaughlin released her work, *Evaluation and Reform: The Elementary and Secondary Education Act of 1965, Title I*. Like other books in the mid-1970s, such as Eugene Bardach's *The Implementation Game: What Happens After a Bill Becomes a Law*, these volumes asked what, if anything, federal officials had learned from the first decade of Great Society program implementation. Their answer: not enough.[14]

Noting the failure of federal policies to achieve stated objectives, researchers in the 1970s asked how much discretion local officials should have over implementation—especially "street-level bureaucrats" who carry out policies day to day. The phrase *street-level bureaucrats* first appeared in Richard Weatherly and Michael Lipsky's critique of the implementation of the Education for All Handicapped Children Act of 1975 (P.L. 94-142). Their article in *Harvard Educational Review*, "Street-Level Bureaucrats and Institutional Innovation: Implementing Special Education Reform," led to Lipsky's book, *Street-Level Bureaucracy: Dilemmas of the Individual in Public Services*.[15]

By the late 1970s, as historians, sociologists, and political scientists continued to study the effects of Johnson-era education programs, the federal education bureaucracy itself came under scrutiny. Critics argued that, just as local constituents used policies to suit their own needs, so did bureaucrats, who sought above all to protect their own jobs. Some asserted that education bureaucrats deliberately failed to achieve policy objectives (or even attempted to "grow" problems they were supposed to solve) in order to demand more resources and more jobs. Scholars asked cynically whether the federal bureaucracy really solved problems or just searched for new problems to bolster its own role.[16]

Kaestle and Smith addressed the issue of federal bureaucratic expansion in their aforementioned article for the *Harvard Educational*

Review. They noted that "the Elementary and Secondary Education Act of 1965 contained only five titles and about a dozen program authorities: revisions in 1968 added new titles and programs, including the Bilingual Education Act. The Education for All Handicapped Children Act passed in 1975, and embellishments to ESEA in 1974 and 1978 eventually enlarged its scope to thirteen titles and over 100 programs. As programs proliferated, so did the bureaucracy, regulations, and interest groups."[17]

Several works have examined the expansion of the federal education bureaucracy, including Lee Sproull, Stephen Weiner, and David Wolf, *Organizing an Anarchy: Belief, Bureaucracy, and Politics in the National Institute of Education*, and Beryl Radin and Willis Hawley, *The Politics of Federal Reorganization: Creating the Department of Education*. These books revealed not only the organizational dynamics of the vast federal education bureaucracy, but also the ways in which it became a magnet for interest-group politics. Radin and Hawley's study of the creation of the Department of Education, for example, showed how debates over the Department mobilized lobbies nationwide.[18]

The historiography on the federal role in education has been shaped dramatically by the rise of education-related interest groups, which often publish their own "histories" of the federal role. Usually stressing the needs or demands of specific groups of students (and often pitting one group against others in a contest for limited funds or a struggle for recognition), interest-group histories advance their own criteria for evaluating the success or failure of federal programs. In so doing, they shape the terms of subsequent legislative debate as well as the interpretive lenses historians use to describe these debates—and the interest groups' own role in them.[19]

In some cases, interest-group histories became influential texts in their respective fields. Examples from the field of bilingual education include Susan Gilbert Schneider's legislative history, *Revolution, Reaction, or Reform: The 1974 Bilingual Education Act*, and James Crawford's book, *Bilingual Education: History, Politics, Theory, and Practice*.[20] Another, more recent history of bilingual education is Guadalupe San Miguel, Jr.'s *Contested Policy: The Rise and Fall of Federal Bilingual Education in the United States, 1960–2001*, which illustrates how diverse interest groups reshaped the debate over federal bilingual education policy over time.[21]

Interest groups constantly jockey for influence within the federal government and within presidential administrations. Of course, some presidents have used these groups to political advantage while others have not. Two well-known histories of education and the presidency are Hugh Davis Graham, *The Uncertain Triumph: Federal Education Policy in the Kennedy and Johnson Years*, and Lawrence McAndrews, *Broken Ground: John F. Kennedy and the Politics of Education*.[22] More recently, McAndrews has written *The Era of Education: The Presidents and the Schools, 1965–2001*. Other studies in this vein include Chester Finn's *Education and the Presidency*, a post-Nixon conservative review, and Terrel Bell's autobiographical narrative, *Thirteenth Man: A Reagan Cabinet Memoir*.[23]

The irony of Bell's tenure as Secretary of Education under President Reagan was that, instead of dismantling the U.S. Department of Education (as he was asked to do), he strengthened the Department considerably. After the release of *A Nation at Risk* in 1983, the Secretary effectively became the monitor of "quality" and "excellence" in the nation's schools. No thoroughly contextualized history of federal education policy under Reagan has yet been published, but a useful contemporary account—starring both Bell and his successor William Bennett—is John Brademas, *The Politics of Education: Conflict and Consensus on Capitol Hill*.[24]

Reagan's education plan, which built on aspects of Nixon's plan, was to roll back the profusion of federal aid-to-education programs that had emerged since the 1960s. His Educational Consolidation and Improvement Act of 1981 (ECIA) drastically reduced the federal education budget, replaced categorical aid with block grants, and called for rapid deregulation. At the same time, Reagan exhorted voters to hold schools "accountable" for student achievement. This combination of defunding, deregulation, and demands for local accountability marked a new era in the federal role—an era that awaits detailed historical analysis.[25]

Among the programs Reagan cut were those supporting racial desegregation in urban areas. Of course, the subject of desegregation has attracted considerable historical attention over time. From Richard Kluger's tome, *Simple Justice: The History of* Brown v. Board of Education *and Black America's Struggle for Equality*, to J. Anthony Lukas's *Common Ground: A Turbulent Decade in the Lives of Three American Families*, histories of federal desegregation politics have frequently blended insights from legal and political history with great skill.[26]

Few have contributed more to this expansive literature than Gary Orfield, whose books, including *The Reconstruction of Southern Education: The Schools and the 1964 Civil Rights Act*, *Must We Bus? Segregated Schools and National Policy*, and *Dismantling Desegregation: The Quiet Reversal of* Brown v. Board of Education, have charted the rise and fall of the federal courts' ambitious crusade for racial justice. Orfield attributes the shifting fortunes of desegregation policies to the evolving composition of the courts and, in turn, the changing priorities of successive presidential administrations. Without vigorous support from the judicial and executive branches, he reasons, desegregation is unlikely to prevail.[27]

Some of the best work on desegregation politics has come from local case studies. Ronald Formisano, *Boston Against Busing: Race, Class, and Ethnicity in the 1960s and 1970s*; David Cecelski, *Along Freedom Road: Hyde County, North Carolina, and the Fate of Black Schools in the South*; Jack Dougherty, *More Than One Struggle: The Evolution of Black School Reform in Milwaukee*; and, more recently, R. Scott Baker, *Paradoxes of Desegregation: African American Struggles for Educational Equity in Charleston, South Carolina, 1926–1972*, all document enduring opposition to integration, not only among whites but also blacks who valued "neighborhood schools."[28]

What these and other works have shown is the remarkably limited power of the federal government to alter the racial composition of the nation's schools. In the absence of widespread political support, desegregation has made little headway, and many have grappled with this issue. James Patterson's *Brown v. Board of Education: A Civil Rights Milestone and Its Troubled Legacy* points out that de jure segregation was much easier to dismantle than de facto segregation, while Mary Ehrlander's *Equal Educational Opportunity: Brown's Elusive Mandate* finds that parents have consistently placed academic achievement above racial desegregation as their highest educational priority.[29]

Kaestle and Smith themselves reflected on the history of desegregation in 1982. "Faced with court-ordered busing in northern cities, the liberal coalition that had spawned the Civil Rights Act and prodded the South split apart," they lamented. "After a decade of confusing evaluation reports, the educational promise of desegregation was no longer clear. The force of black and liberal public opinion had diminished; substantial elements of the black community believed that there were other approaches to equal opportunity and racial justice than school desegregation."[30] Despite the symbolic meaning of *Brown*, they concluded, no branch of the federal government had unambiguously endorsed racial desegregation as the *sine qua non* of equal educational opportunity.[31]

For Kaestle and Smith, the failure of desegregation signified a larger failure of the federal role. "The clear national vision that linked racial desegregation with educational equity and generated *Brown* and its aftermath has dissipated," they saw. "We now live in a country where huge sections of central cities are almost all-black, where desegregation is often impossible even if district borders are ignored, and where many of the central city schools do not even remotely resemble suburban schools in . . . quality. So long as the federal government does not take clear and substantive action to meet the needs of inner cities, it forfeits any claim to being a nation with racial justice."[32]

Many believe that racial desegregation, if accomplished, could improve academic achievement, but a causal link between desegregation and achievement has been hard to prove. The factors most powerfully correlated with achievement seem to have more to do with family income than racial balance, a finding that has been virtually unshakable since the mid-1960s.[33] Even a link between federal aid and student achievement has remained unclear: as Kaestle and Smith found after reviewing data from the National Assessment of Educational Progress (NAEP) in 1982, it was hard to discern whether Title I aid, spent on an array of programs, actually *caused* gains in achievement for particular students.[34]

Since the 1980s, the issue of academic achievement has gained prominence, and the so-called standards movement has begun to attract historical interest. Two accessible histories of the standards movement are Diane Ravitch, *National Standards in American Education: A Citizen's Guide*, and John F. Jennings, *Why National Standards and Tests? Politics and the Quest for Better Schools*, both of which trace the origins of the standards movement back to the early implementation of the ESEA in the 1960s and highlight the many accountability initiatives that began at the state level in the 1970s and 1980s.[35]

States have, of course, played a leading role in the standards movement. In 1989, Arkansas governor Bill Clinton and Tennessee governor Lamar Alexander held a meeting in Charlottesville, Virginia, at which they (and forty-seven other governors) drafted a compact (1) to focus on setting academic performance objectives in every state, and (2) to frame these objectives as "national priorities." After this conference, newly elected President George H. W. Bush made these goals part of his America 2000 Education Initiative.[36] Later, the Clinton administration picked up where Bush left off, emphasizing state-level standards in its Improving America's Schools Act of 1994 (which reauthorized the ESEA).[37]

In the 1990s, various scholars began to examine the standards movement as it was playing out in state and federal education policies.[38] Helen F. Ladd, ed., *Holding Schools Accountable: Performance-Based Reform in Education*, generally supported the movement, while Margaret E. Goertz, in "Redefining Government Roles in an Era of Standards-Based Reform," published in *Phi Delta Kappan*, expressed more ambivalence.[39] Both scholars, however, showed how standards-oriented centralization at the *state* level had converged with a similar process of standards-oriented centralization at the *federal* level over time—a process apparently supported by leaders of both political parties.[40]

Much recent debate over standardized testing has focused on the provisions of the bipartisan No Child Left Behind Act of 2002. A thorough legislative history is Elizabeth DeBray's *Politics, Ideology, and Education: Federal Policy during the Clinton and Bush Administrations*, which shows how a growing number of interest groups and think tanks displaced traditional education groups and congressional staffers in writing the new law. In the same vein, Patrick McGuinn's *No Child Left Behind and the Transformation of Federal Education Policy, 1965–2005* examines how this "centrist" law reflects a series of political compromises that were necessary in the context of party parity in the 106th and 107th congresses.[41]

While the long-term consequences of the No Child Left Behind Act remain to be seen, historical debate over "the federal role" in the public schools is certain to continue. Specifically, the *effectiveness* of the federal role is likely to be assessed by the degree to which schools receiving federal grants raise achievement for all students—as measured by scores on nationally standardized tests. How well schools will meet this challenge is arguably the most urgent question facing

272   ADAM R. NELSON

American education today. Can historians shed any light on the prospects for success?

Of all the responsibilities of the federal government *vis à vis* education, that of funding research on "best practices" has perhaps the most contentious history, and this history has itself attracted growing interest. Carl Kaestle's essay "The Awful Reputation of Education Research," published in the *Educational Researcher*, launched the discussion, while Ellen Condliffe Lagemann's *An Elusive Science: The Troubling History of Education Research*, and David Labaree's *The Trouble With Ed Schools*, fueled the debate. All three historians offered rather disheartening interpretations of the politics of federally sponsored educational research.[42]

The federal government has, of course, sponsored thousands of studies to assess the effectiveness of federal education programs. Yet, despite various reconfigurations of the federal education research agency—from the National Institute of Education (NIE) to the Office of Educational Research and Improvement (OERI) to the current Institute for Education Sciences (IES)—the impression persists that large-scale grants-in-aid have not dramatically improved the quality (or the *equality*) of the nation's schools. To deal with this poor record, some have called for a new type of research agency; others have called for a new type of research.

The ascendant "gold standard" in federally-sponsored educational research is the randomized field trial, which, like other forms of scientific analysis, rigorously controls for local variations or demographic idiosyncrasies to produce statistically valid (and thus generalizable) results. Books such as Richard Shavelson and Lisa Towne's edited volume, *Scientific Research in Education*, while not entirely dismissing qualitative methodologies, nonetheless emphasizes quantitative techniques, particularly those based on econometric or psychometric models. This emphasis has sidelined high-quality research in fields like history—research that highlights the ways in which local factors often unsettle even the best-designed policies.[43]

The lack of federal support for historical research in education has led at least one state, New York, to attempt to fill this void. In 2003, with a sizable grant from the New York Community Trust-Wallace Foundation Special Projects Fund, the New York State Archives began an innovative "States' Impact on Federal Education Policy Project." This effort puts historians in touch with the archival materials they need to study the evolution of state-federal relations in educational policy. Project advisors include Gordon Ambach, former state commissioner of education in New York; John F. Jennings, former general counsel to the House Committee on Education and Labor; and Carl Kaestle.[44]

What historians bring to education research is, of course, the long view, the wide perspective, the comprehensive outlook that is essential for successful policy. Historians can help policymakers see, for example, the broad effects of a half-century of large-scale federal involvement in the nation's schools: the anemic commitment to racial integration, the persistent inequality in school resources, the enduring gaps in pupil achievement, and the growing appeals for vouchers that steer public funds to private schools. While federal policies alone did not *cause* these results, neither did federal policies (ostensibly intended to equalize educational opportunity) avert them. Historians can help explain why.

Twenty-five years ago, Kaestle and Smith were not sanguine about the prospects of solving what appeared to be intractable problems in the nation's schools. "The lack of confidence in public schools, the decline in attractiveness of the teaching profession, the desperate condition of inner-city schools, and continuing separation in the schooling of blacks and whites and rich and poor cannot be met by programs and policies that operate on the periphery of the regular school program," they wrote. "If federal involvement is to influence these issues, the strategies and authority of the federal government would have to increase in a fashion similar to the long-range historical increase in state authority."[45]

History suggests, however, that increasing federal authority over state and local authority in education poses significant challenges. Even if greater federal authority over education were possible, would it be desirable? Historians disagree. Acknowledging the complexity, the contradictions, and the many unintended consequences of federal aid to education in the past, even Kaestle and Smith asked themselves if greater centralization at the federal level was a good idea. Sensing that centralization at the federal level was perhaps inevitable, they offered some advice. "Our prescription to federal policymakers," they concluded, "is simple: carefully choose educational issues that focus on basic constitutional rights and social justice; use funding and regulations to integrate these goals into the central, day-to-day activities of local schools across the nation; and construct programs that will foster local decision-making while transcending

274 ADAM R. NELSON

local variation."[46] This was sound advice, but, historically speaking, it was hardly simple.

NOTES

1. Carl F. Kaestle and Marshall S. Smith, "The Federal Role in Elementary and Secondary Education, 1940–1980," *Harvard Educational Review* 52 (November 1982): 386.

2. Kaestle and Smith, "The Federal Role in Elementary and Secondary Education," 385.

3. An exception is Adam R. Nelson, *The Elusive Ideal: Equal Educational Opportunity and the Federal Role in Boston's Public Schools, 1950–1985* (Chicago: University of Chicago Press, 2005).

4. For more on the federal role in education before World War II, see David Tyack, Thomas James, and Aaron Benavot, *Law and the Shaping of Public Education, 1785–1954* (Madison: University of Wisconsin Press, 1987). For more on the Northwest Ordinances, see Paul H. Mattingly and Edward W. Stevens, Jr., eds., *"—schools and the means of education shall forever be encouraged": A History of Education in the Old Northwest, 1787–1880* (Athens, Ohio: Ohio University Libraries, 1987). For more on the Morrill Act, see Earle Ross, *Democracy's College: The Land-Grant Movement in the Formative State* (New York: Arno, 1969), and Frederick B. Mumford, *The Land Grant College Movement* (Columbia, MO: University of Missouri College of Agriculture and Agricultural Experiment Station, 1940). For more on the Freedmen's Bureau, see Donald G. Nieman, ed. *The Freedmen's Bureau and Black Freedom* (New York: Garland, 1994). See also Robert C. Morris, "Educational Reconstruction," in *The Facts of Reconstruction: Essays in Honor of John Hope Franklin*, ed. Eric Anderson and Alfred A. Moss, Jr. (Baton Rouge: Louisiana State University Press, 1991), 141–66; James D. Anderson, *The Education of Blacks in the South, 1860–1935* (Chapel Hill: University of North Carolina Press, 1988), 4–32; and Michael Goldhaber, "A Mission Unfulfilled: Freedmen's Education in North Carolina, 1865–1870," *Journal of Negro History* 77 (Spring 1992): 199–210.

5. Gordon C. Lee, *The Struggle for Federal Aid. First Phase: A History of the Attempts to Obtain Federal Aid for the Common Schools, 1870–1890* (New York: Teachers College, 1949); Herbert M. Kliebard, *Schooled to Work: Vocationalism and the American Curriculum, 1876–1946* (New York: Teachers College, 1999). For more on Smith-Hughes, see Sol Cohen, "The Industrial Education Movement, 1906–1917," *American Quarterly* 20 (1968): 95–110; and Herbert M. Kliebard, *The Struggle for the American Curriculum, 1893–1958*, 3rd ed. (New York: Routledge, 2004). For more on the federal role before World War II, see Hollis Allen, *The Federal Government and Education* (New York: McGraw-Hill, 1950), and Frank Munger and Richard Fenno, *National Politics and Federal Aid to Education* (Syracuse: Syracuse University Press, 1962).

6. Gilbert E. Smith, *The Limits of Reform: Politics and Federal Aid to Education, 1937–1950* (New York: Garland, 1982); Richard Reiman, *The New Deal and American Youth: Ideas and Ideals in a Depression Decade* (Athens: University of Georgia Press, 1992). For more on federalism during the New Deal, see James T. Patterson, *The New Deal and the States: Federalism in Transition* (Princeton: Princeton University Press, 1969), and *Congressional Conservatism and the New Deal: The Growth of the Conservative Coalition in Congress, 1933–1939* (Lexington: University of Kentucky

Press, 1967). See also Charles A. Quattlebaum, *Federal Aid to Elementary and Secondary Education: An Analytic Study of the Issue, Its Background, and Relevant Legislative Proposals* (Chicago: Public Administration Service, 1948).

7. Barbara Barksdale Clowse, *Brainpower for the Cold War: The Sputnik Crisis and National Defense Education Act of 1958* (Westport, CT: Greenwood, 1981). See also John L. Rudolph, *Scientists in the Classroom: The Cold War Reconstruction of American Science Education* (New York: Palgrave Macmillan, 2002).

8. Robert Bendiner, *Obstacle Course on Capitol Hill* (New York: McGraw Hill, 1964); Virgil Blum, *Freedom in Education: Federal Aid for All Children* (Garden City, NY: Doubleday, 1965); Sidney Tiedt, *The Role of the Federal Government in Education* (New York: Oxford University Press, 1966).

9. Stephen Bailey and Edith Mosher, *ESEA: The Office of Education Administers a Law* (Syracuse: Syracuse University Press, 1968); Eugene Eidenberg and Roy Morey, *An Act of Congress: The Legislative Process and the Making of Education Policy* (New York: Norton, 1969). For more on fading concerns about "federal control" in the wake of growing social-welfare programs, see Linda Bennett and Stephen Bennett, *Living With Leviathan: Americans Coming to Terms with Big Government* (Lawrence, KS: University of Kansas Press, 1990). The Bennetts contend that, over time, Americans have become "habituated" to federal aid. They write: "Americans have come to accept the wide range of functions and services the national government provides, and their suspicion of government's power stops far short of their demands for its withdrawal. In relations between citizen and government, many Americans have come to see themselves more as beneficiaries of its largess than as potential victims of its power" (*xi*). They conclude that, "Whatever Americans may say about big government in the abstract, they have a healthy and continuing appetite for the programs that make up the positive state" (*xiii*).

10. Julie Roy Jeffrey, *Education for Children of the Poor: A Study of the Origins and Implementation of the Elementary and Secondary Education Act of 1965* (Columbus: Ohio State University Press, 1978). For more on education and poverty relief, see James S. Coleman et al., *Equality of Educational Opportunity* (Washington: U.S. Dept. of Health, Education, and Welfare, Office of Education, 1966). See also James Patterson *America's Struggle Against Poverty in the Twentieth Century* (Cambridge, MA: Harvard University Press, 2000), and Michael Katz, ed., *The "Underclass" Debate: Views From History* (Princeton: Princeton University Press, 1993). For more on the ESEA, see Frederick Mosteller and Daniel P. Moynihan, eds., *On Equality of Educational Opportunity* (New York: Vintage, 1972); Jerome T. Murphy, *State Education Agencies and Discretionary Funds: Grease the Squeaky Wheel* (Lexington, MA: Lexington Books, 1974); Harvey Kantor, "Education, Social Reform, and the State: ESEA and Federal Education Policy in the 1960s," *American Journal of Education* 100 (November 1991): 47–83; Samuel Halperin, "ESEA Comes of Age: Some Historical Reflections," *Educational Leadership* 36 (February 1979): 349–53; and Patrick McGuinn and Frederick Hess, "Freedom from Ignorance? The Great Society and the Evolution of the Elementary and Secondary Education Act of 1965," in *The Great Society and the High Tide of Liberalism*, ed. Sidney Milkis and Jerome Mileur (Amherst: University of Massachusetts Press, 2005).

11. Maris Vinovskis, *The Birth of Head Start: Preschool Education Policies in the Kennedy and Johnson Years* (Chicago: University of Chicago Press, 2005). For more on the history of Head Start, see Maris Vinovskis, *History and Educational Policymaking* (New Haven: Yale University Press, 1999), and Maris Vinovskis, "Do Federal Compensatory Education Programs Really Work? A Brief Historical Analysis of Title I and Head Start," *American Journal of Education* 107 (May 1999): 187–209. See also

Jeanne Elsworth and Linda J. Ames, eds., *Critical Perspectives on Project Head Start: Revisioning the Hope and Challenge* (Albany: State University of New York Press, 1998), and Kay Mills, *Something Better for My Children: The History and People of Head Start* (New York: Dutton, 1998).

12. Jerome T. Murphy, "Title I of ESEA: The Politics of Implementing Federal Educational Reform," *Harvard Educational Review* (1971): 35–63; Ruby Martin and Phyllis McClure, *Title I of ESEA: Is It Helping Poor Children?* (Washington: Washington Research Project of the Southern Center for Studies in Public Policy and the NAACP Legal Defense and Educational Fund, Inc., 1969). For more on the implementation of Title I, see Theodore Sky, "Concentration under Title I of the Elementary and Secondary Education Act," *Journal of Law and Education* 1, vol. 2 (1972): 171–211; Jerome T. Murphy, "The Education Bureaucracies Implement a Novel Policy: The Politics of Title I of ESEA, 1965–1972," in *Policy and Politics in America: Six Case Studies*, ed. Alan P. Sindler (Boston: Little, Brown, and Company, 1973), 160–98; and Gary Natriello and Edward McDill, "Title I: From Funding Mechanism to Educational Program," in *Hard Work for Good Schools: Facts Not Fads in Title I Reform*, ed. Gary Orfield and Elizabeth DeBray (Cambridge, MA: Harvard University, Civil Rights Project, 1999).

13. Edward Wynne's *Politics of School Accountability: Public Information About Public Schools* (Berkeley: McCutchan, 1972); Jerome T. Murphy and David Cohen, "Accountability in Education: The Michigan Experience," *The Public Interest* 36 (Spring 1974): 53–81. For more on early accountability measures, see Walter Haney and George Madaus, "Making Sense of the Competency Testing Movement," *Harvard Educational Review* 48 (Winter 1978): 462–84; Richard M. Jaeger and Carol K. Tittle, eds. *Minimum Competency Achievement Testing: Motives, Models, Measures, and Consequences* (Berkeley, CA: McCutchan, 1980); and Myron Atkin and Ernest House, "The Federal Role in Curriculum Development, 1950–1980," *Educational Evaluation and Policy Analysis* 3 (March 1981): 5–36.

14. Jeffrey Pressman and Aaron Wildavsky, *Implementation: How Great Expectations in Washington are Dashed in Oakland; or Why It's Amazing That Federal Programs Work At All* (Berkeley: University of California Press, 1973); Milbrey W. McLaughlin, *Evaluation and Reform: The Elementary and Secondary Education Act of 1965, Title I* (Cambridge, MA: Ballinger, 1975); Eugene Bardach, *The Implementation Game: What Happens After a Bill Becomes a Law* (Cambridge, MA: MIT Press, 1977). See also Milbrey W. McLaughlin, "Implementation of ESEA Title I: A Problem of Compliance," *Teachers College Record* 77 (Summer 1976): 397–415.

15. Richard Weatherly and Michael Lipsky, "Street-Level Bureaucrats and Institutional Innovation: Implementing Special Education Reform," *Harvard Educational Review* 47 (Summer 1977): 89–119; Michael Lipsky, *Street-Level Bureaucracy: Dilemmas of the Individual in Public Services* (New York: Russell Sage Foundation, 1980).

16. For more on bureaucratic expansion in the welfare state, see James T. Bennett and Manuel H. Johnson, *The Political Economy of Federal Government Growth, 1959–1978* (College Station, TX.: Texas A&M University Center for Education, June 1980). "If bureaucratic [self-preservation] is the motivating factor, it is axiomatic that no crises can ever be solved; if a crisis disappears, the justification for the agency, its employees, and its appropriations would also vanish. Bureaucratic entrepreneurship requires that funding must increase over time, and for this to occur, problems must multiply as well. There is no incentive for the careful management of the taxpayers' money, for if an agency does not spend all its money in one budget year, additional appropriations in the following year may be in jeopardy. Bureaucratic failure is, from the perspective of the bureaucrat, success" (117).

17. Kaestle and Smith, "The Federal Role in Elementary and Secondary Education," 404. For more on the accretion of federal education programs, see Jackie Kimbrough and Paul Hill, *The Aggregate Effects of Federal Education Programs* (Santa Monica, CA: Rand Corporation, 1981). See also David Tyack and Larry Cuban, *Tinkering Toward Utopia: A Century of Public School Reform* (Cambridge, MA: Harvard University Press, 1995).

18. Lee Sproull, Stephen Weiner, and David Wolf, *Organizing an Anarchy: Belief, Bureaucracy, and Politics in the National Institute of Education* (Chicago: University of Chicago Press, 1978); Beryl Radin and Willis Hawley, *The Politics of Federal Reorganization: Creating the Department of Education* (New York: Pergamon, 1988).

19. See Stephen Bailey, *Education Interest Groups in the Nation's Capital* (Washington, DC: American Council on Education, 1975); and Paul E. Peterson and Barry G. Rabe, "The Role of Interest Groups in the Formation of Educational Policy: Past Practice and Future Trends," *Teachers College Record* 85 (Fall 1983): 708–29.

20. Susan Gilbert Schneider, *Revolution, Reaction, or Reform: The 1974 Bilingual Education Act* (New York: Las Americas, 1976); James Crawford, *Bilingual Education: History, Politics, Theory, and Practice* (Los Angeles: Bilingual Educational Services, 1989). For more on the history of bilingual education debates, see Kenji Hakuta, *The Mirror of Language: The Debate on Bilingualism* (New York: Basic Books, 1986). See also Herbert Teitlebaum and Richard J. Hiller, "Bilingual Education: The Legal Mandate," *Harvard Educational Review* (May 1977): 138–70; Gaynor Cohen, "Alliance and Conflict Among Mexican Americans," *Ethnic and Racial Studies* 5(2): 175–95; Isaura Santiago Santiago, "*Aspira v. Board of Education* Revisited," *American Journal of Education* (November 1986): 149–99; Joseph Watras, "Bilingual Education and the Campaign for Federal Aid," *American Educational History Journal* 27 (2000): 81–87; and Gareth Davies, "The Great Society After Johnson: The Case of Bilingual Education," *Journal of American History* 88 (March 2002): 1405–29.

21. Guadalupe San Miguel, Jr., *Contested Policy: The Rise and Fall of Federal Bilingual Education in the United States, 1960–2001* (Denton, TX: University of North Texas Press, 2004). For examples from the field of special education, see Erwin Levine and Elizabeth Wexler, *P.L. 94-142: An Act of Congress* (New York: Macmillan, 1981), and Joseph Ballard, Bruce Ramirez, and Frederick Weintraub, *Special Education in America: Its Legal and Governmental Foundations* (Reston, VA: Council for Exceptional Children, 1982). In the 1970s and 1980s, scholars used the phrase *iron triangle* to characterize the close relationships that existed between interest groups, congressional subcommittees, and regulatory agencies in many policy arenas, including education.

22. Hugh Davis Graham, *The Uncertain Triumph: Federal Education Policy in the Kennedy and Johnson Years* (Chapel Hill: University of North Carolina Press, 1984); Lawrence McAndrews, *Broken Ground: John F. Kennedy and the Politics of Education* (New York: Garland, 1991). See also James L. Sundquist, *Politics and Policy: The Eisenhower, Kennedy, and Johnson Years* (Washington, DC: Brookings Institution, 1968).

23. Lawrence J. McAndrews, *The Era of Education: The Presidents and the Schools, 1965–2001* (Urbana, IL: University of Illinois Press, 2006); Chester E. Finn, *Education and the Presidency* (Lexington, MA: Lexington Books, 1977); Terrel H. Bell, *Thirteenth Man: A Reagan Cabinet Memoir* (New York: Free Press, 1988). Also useful on the topic of the executive role is Frederick Hess and Patrick McGuinn's survey in "Seeking the Mantle of Opportunity: Presidential Politics and the Educational Metaphor, 1964–2000," *Educational Policy* 16 (March, 2002): 72–95.

24. National Commission on Excellence in Education, *A Nation at Risk: The Imperative for Educational Reform. A Report to the Nation and the Secretary of Education* (Washington, DC: United States Department of Education, 1983); John Brademas, *The Politics of Education: Conflict and Consensus on Capitol Hill* (Norman, OK: University of Oklahoma Press, 1987). For more on education policy in the Reagan years, see David Cohen, "Policy and Organization: The Impact of State and Federal Policy on School Governance," *Harvard Educational Review* (1982): 474–99; Jeffrey Henig, *Public Policy and Federalism* (New York: St. Martins, 1985); Richard Jung and Michael Kirst, "Beyond Mutual Adaptation, Into the Bully Pulpit: Recent Research on the Federal Role in Education," *Educational Administration Quarterly* 22 (Summer 1986): 80–109; and, Bruce Cooper and Denis Doyle, eds. *Federal Aid to the Disadvantaged: What Future for Chapter 1?* (New York: Falmer, 1988). For a broader history shaped by the politics of the Reagan era, see Diane Ravitch, *The Troubled Crusade: American Education, 1945–1980* (New York: Basic Books, 1983).
25. See Rosemary Salomone, *Equal Education Under Law: Legal Rights and Federal Policy in the Post-Brown Era* (New York: St. Martin's, 1986).
26. Richard Kluger, *Simple Justice: The History of* Brown v. Board of Education *and Black America's Struggle for Equality* (New York: Knopf, 1976); J. Anthony Lukas, *Common Ground: A Turbulent Decade in the Lives of Three American Families* (New York: Knopf, 1985).
27. Gary Orfield, *The Reconstruction of Southern Education: The Schools and the 1964 Civil Rights Act* (New York: Wiley-Interscience, 1968), *Must We Bus? Segregated Schools and National Policy* (Washington: Brookings Institution, 1978), and *Dismantling Desegregation: The Quiet Reversal of* Brown v. Board of Education (New York: New Press, 1996). See also Beryl A. Radin, *Implementation, Change, and the Federal Bureaucracy: School Desegregation Policy in H.E.W., 1964–1968* (New York: Teachers College Press, 1977).
28. Ronald P. Formisano, *Boston Against Busing: Race, Class, and Ethnicity in the 1960s and 1970s* (Chapel Hill: University of North Carolina Press, 1991); David S. Cecelski, *Along Freedom Road: Hyde County, North Carolina, and the Fate of Black Schools in the South* (Chapel Hill: University of North Carolina Press, 1994); Jack Dougherty, *More Than One Struggle: The Evolution of Black School Reform in Milwaukee* (Chapel Hill: University of North Carolina Press, 2004); R. Scott Baker, *Paradoxes of Desegregation: African American Struggles for Educational Equity in Charleston, South Carolina, 1926–1972* (Columbia, SC: University of South Carolina Press, 2006). Hundreds of books and articles have addressed the subject of racial desegregation. See, for example, James Duran *A Moderate Among Extremists: Dwight D. Eisenhower and the School Desegregation Crisis* (Chicago: Nelson-Hall, 1981); Tony Freyer *The Little Rock Crisis: A Constitutional Interpretation* (Westport, CT: Greenwood, 1984); Jennifer Hochschild, *The New American Dilemma: Liberal Democracy and School Desegregation* (New Haven: Yale University Press, 1984); J. Brian Sheehan, *The Boston School Integration Dispute: Social Change and Legal Maneuvers* (New York: Columbia University Press, 1984); Daniel J. Monti, *A Semblance of Justice: School Desegregation and the Pursuit of Order in Urban America* (Columbia, MO.: University of Missouri Press, 1985); and Susan E. Eaton, *The Other Boston Busing Story: What's Won and Lost Across the Boundary Line* (New Haven: Yale University Press, 2001). See also Susan M. McGrath, "From Tokenism to Community Control: Political Symbolism in the Desegregation of Atlanta's Public Schools, 1961–1973," *Georgia Historical Quarterly* 79 (Winter 1995): 842–72; R. David Riddle, "Race and Reaction in Warren, Michigan, 1971–1974: *Bradley v. Milliken* and the Cross-District Busing Controversy," *Michigan Historical Review* 26 (Summer 2000): 1–49; and

Tom I. Romero II, "Our Selma Is Here: The Political and Legal Struggle for Educational Equality in Denver, Colorado, and Multiracial Conundrums in American Jurisprudence," *Seattle Journal for Social Justice* 3 (Fall/Winter 2004): 73–142.
29. James T. Patterson, *Brown v. Board of Education: A Civil Rights Milestone and Its Troubled Legacy* (New York: Oxford University Press, 2001); Mary Ehrlander, *Equal Educational Opportunity: Brown's Elusive Mandate* (New York: LFB Scholarly Pub., 2002). On the contest between academic achievement and racial desegregation, see also Jennifer Hochschild and Nathan Scovronick, *The American Dream and the Public Schools* (New York: Oxford University Press, 2003).
30. Kaestle and Smith, "The Federal Role in Elementary and Secondary Education," 403.
31. On "community control" in this period, a classic study remains Daniel P. Moynihan, *Maximum Feasible Misunderstanding: Community Action in the War on Poverty* (New York: Free Press, 1969).
32. Kaestle and Smith, "The Federal Role in Elementary and Secondary Education," 404. Later decades have brought anniversary volumes on *Brown*. See, for example, Raymond Wolters, *The Burden of Brown: Thirty Years of School Desegregation* (Knoxville: University of Tennessee Press, 1984), and Deborah L. Rhode and Charles J. Ogletree, Jr., Brown *at Fifty: The Unfinished Legacy: A Collection of Essays* (Chicago: American Bar Association, 2004). See also Michael Fultz, "The Displacement of Black Educators Post-*Brown*: An Overview and Analysis," *History of Education Quarterly*, 44 (1): 11–45.
33. See Coleman, op. cit., *Equality of Educational Opportunity* (1966).
34. Kaestle and Smith, "The Federal Role in Elementary and Secondary Education," 396–400.
35. Diane Ravitch, *National Standards in American Education: A Citizen's Guide* (Washington: Brookings Institution, 1995); John F. Jennings, *Why National Standards and Tests? Politics and the Quest for Better Schools* (Thousand Oaks, CA: Sage Publications, 1998). See also Susan H. Fuhrman, ed., *From the Capitol to the Classroom: Standards-Based Reform in the States* (Chicago: National Society for the Study of Education, 2001).
36. For more on America 2000, see John F. Jennings's edited series, *National Issues in Education*, particularly the volume *The Past is Prologue* (Washington: Institute for Educational Leadership, 1993). See also Frederick Wirt and Michael Kirst, *The Political Dynamics of American Education* (Richmond, CA: McCutchan, 1997).
37. Many scholars have addressed the impact of federal aid on state education policy. See Diane Massell and Susan Fuhrman with Michael Kirst, *Ten Years of State Education Reform, 1983–1993: Overview with Four Case Studies* (New Brunswick, NJ: Consortium for Policy Research in Education, 1994), and Thomas B. Timar, "The Institutional Role of State Education Departments: A Historical Perspective," *American Journal of Education* 105 (May 1997): 231–60.
38. For more on the Clinton program, *Goals 2000*, see Robert B. Schwartz and Marion A. Robinson, "Goals 2000 and the Standards Movement" in *Brookings Papers on Education Policy: 2000*, ed. Diane Ravitch (Washington: Brookings Institution, 2000), 173–214.
39. Helen F. Ladd, ed., *Holding Schools Accountable: Performance-Based Reform in Education* (Washington: Brookings Institution, 1996); Margaret E. Goertz, "Redefining Government Roles in an Era of Standards-Based Reform," *Phi Delta Kappan* 83 (September 2001), 62–66.
40. For criticism of the standards movement and its effect on curricular priorities and classroom pedagogy, see Linda M. McNeil, *Contradictions of School Reform: The Educational Costs of Standardized Testing* (New York: Routledge, 2000); Martin Carnoy,

280   ADAM R. NELSON

Richard Elmore, and Leslie Santee Siskin, eds., *The New Accountability: High Schools and High Stakes Testing* (New York: Routledge Falmer, 2003); James Spillane, *Standards Deviation: How Schools Misunderstand Education Policy* (Cambridge, MA: Harvard University Press, 2004); Kevin Kosar, *Failing Grades: The Federal Politics of Education Standards* (Boulder: L. Rienner, 2005); Alfie Kohn, *The Case Against Standardized Testing: Raising the Scores, Ruining the Schools* (Portsmouth, NH: Heinemann, 2000); and M. Gail Jones, *The Unintended Consequences of High-Stakes Testing* (Lanham, MD: Rowman & Littlefield, 2003).

41. Elizabeth H. DeBray, *Politics, Ideology, and Education: Federal Policy During the Clinton and Bush Administrations* (New York: Teachers College Press, 2006); Patrick J. McGuinn's *No Child Left Behind and the Transformation of Federal Education Policy, 1965–2005* (Lawrence, KS: University Press of Kansas, 2006). For criticism of the No Child Left Behind Act, see Deborah Meier and George Wood, eds. *Many Children Left Behind: How the No Child Left Behind Act Is Damaging Our Children and Our Schools* (Boston: Beacon, 2004); Joan Herman and Edward Haertel, eds., *Uses and Misuses of Data for Educational Accountability and Improvement* (Chicago: National Society for the Study of Education, 2005); and Susan Fuhrman and Richard Elmore, eds., *Redesigning Accountability Systems for Education* (New York: Teachers College Press, 2005).

42. Carl F. Kaestle, "The Awful Reputation of Education Research," *Educational Researcher* 22 (January/February 1993): 23, 26–31; Ellen Condliffe Lagemann, *An Elusive Science: The Troubling History of Education Research* (Chicago: University of Chicago Press, 2000); David F. Labaree, *The Trouble With Ed Schools* (New Haven: Yale University Press, 2004). For more on the "awful reputation" of education research, see Jurgen Herbst, *And Sadly Teach: Teacher Education and Professionalization in American Culture* (Madison: University of Wisconsin Press, 1989); National Research Council, *Improving Student Learning: A Strategic Plan for Education Research and Its Utilization* (Washington: National Academy Press, 1999); Maris Vinovskis, "The Federal Role in Educational Research and Development," in *Brookings Papers on Education Policy: 2000*, ed. Diane Ravitch (Washington: Brookings Institution, 2000), 359–96; Frederick Mosteller and Robert Boruch, *Evidence Matters: Randomized Trials in Education Research* (Washington: Brookings Institution, 2002); and David Berliner, "Educational Research: The Hardest Science of All," *Educational Researcher* 31 (November 2002): 18–21.

43. Richard Shavelson and Lisa Towne, eds., *Scientific Research in Education* (Washington: National Academy Press, 2002).

44. For this web-based project, see New York State Archives, "States' Impact on Federal Education Policy" at http://www.sifepp.nysed.gov/edindex.shtml. Also associated with the States' Impact on Federal Education Policy Project is Christopher Cross, author of *Political Education: National Policy Comes of Age* (New York: Teachers College Press, 2004).

45. Kaestle and Smith, "The Federal Role in Elementary and Secondary Education," 407.

46. Kaestle and Smith, "The Federal Role in Elementary and Secondary Education," 408.

Epilogue

New Directions in the History of Education

William J. Reese and John L. Rury

As suggested earlier, the essays in this volume are testimony to the wide scope of the history of American education, and the diversity of topics and perspectives it has grown to embrace. They have provided ample evidence of the field's evolution over a period of several decades following the heyday of the revisionists. It also can be said, on the other hand, that they consider a range of rather traditional topical domains within the history of American education. In this respect it is important to acknowledge some of the emerging areas of scholarly activity that may shape the field in the future.

Over the past several decades, research on a number of relatively new topical issues has developed as the larger field has continued to evolve. Some of these studies have followed somewhat broader trends in American social history, such as a growing interest in the experiences of particular groups that had previously received rather little attention from historians. Themes of exploitation, discrimination, and exclusion that characterized the revisionist era certainly have been evident in this new body of work, but many of these studies have been informed by such new interpretive perspectives as colonialism and resistance theory. While much of this research is still evolving, and its influence is just beginning to be felt, there can be little question about its importance.

As mentioned in the book's introduction, among the most prominent of these new focal points has been the history of education for American Indians, a topic that received very little attention from scholars prior to the revisionist era. One of the early works on the topic, published by Margaret Szasz in the mid-1970s, focused on the era of tribal control following the decline of the infamous boarding schools that had set the tome for American Indian education since the latter nineteenth century. While there had been some attention given to the assimilationist predisposition of the boarding schools, dreadfully expressed in Captain Henry Pratt's oft-quoted intention to "kill the Indian and save the man," it was not until the 1990s that an outpouring of scholarship began to paint a more compelling picture of the period. As suggested earlier, among the most significant of these works was David Wallace Adam's *Education for Extinction*—published in 1996 and recipient of a number of awards—but there have been a number of others as well. Much of this new work has been done by American Indian scholars, as they have started to explore the role of schools and education in the experiences of their forbearers.[1] This is an especially promising trend, one that can help to considerably broaden the diversity of perspectives within the field.

The history of education for Chicano/Latino and Asian children are topics that also have gained greater attention in recent decades. Again, relatively little consideration was afforded these areas during the revisionist era, but scholarship on them has flourished more recently. Michael Olneck has touched upon this body of work in his chapter on immigration and education. It was during the 1980s and 1990s that the experiences of these groups became topics of interest to historians, and especially a new generation of Latino and Asian American scholars. Guadalupe San Miguel, Eileen Tamura, and George L. Sanchez published prize-winning studies of Chicano and Japanese youth and the struggle for equality in education, and a number of other scholars contributed research on yet other questions related to these issues.[2] It is doubtless fair to say that a good deal more research on the educational experiences of these and other understudied groups in American history will appear in the years ahead.

Yet an additional topical area in the larger field that bears mention is the history of teachers and teaching. This too was an area of research that received relatively little attention during the revisionist era. A collection of essays edited by Donald Warren in 1989, however, signaled a new interest in the issue across a broad spectrum of scholars. Published under the auspices of AERA, this book proved to be a catalyst for additional work on this topic, and led to a steady stream of new work dealing with teachers and teaching in American history. Some of this research overlapped with writing on the role of women in the history of education, since women have largely dominated the teaching field in the United States since the latter nineteenth century. But much of it has been concerned with identifying the ways that teaching evolved as a profession, and how certain aspects of it have remained more or less constant up to the present. An early example was Larry Cuban's influential examination of teaching practices throughout the twentieth century, *How Teachers Taught*, which suggested that teaching practices may not have changed nearly as much as previous works had suggested. Other studies focused on the struggles that teachers faced in their daily professional and personal lives. During the 1990s, Kate Rousmaniere and Kathleen Weiler contributed well-received studies of women teachers in contexts as diverse as New York City and rural California. Other historians have provided similar glimpses of teachers' experiences in a number of additional settings. Recently, James Fraser has crafted a compelling synthesis of this body of work, providing a panoramic view of teacher education in the United States, noting key trends and points of discontinuity in its development.[3] No doubt this work will find a ready audience and stimulate additional lines of research.

A final area of research that has flourished in recent years has been the history of educational reform. In certain respects, of course, it is possible to say that this is a topic that has long been of interest to educational historians. Reform, after all, is often just another way of denoting educational change, and almost all historians write about one variety of change or another. But reform usually is a process of self-conscious change, typically under the auspices of a particular understanding or theory of education or human development and institutional power, and with identifiable goals in view. Lawrence Cremin's prize-winning study of Progressive education, *The Transformation of the School*, was undoubtedly the most celebrated study of reform published during the early revisionist period—one that wielded broad influence on subsequent generations of historians.

Cremin offered a generally sympathetic account of progressive educators, depicting them as humanitarians who sought in one way or
another to liberate children from the grip of traditional modes of
instruction and classroom discipline. On the other hand, Michael
Katz's study of nineteenth century change, *The Irony of Early School
Reform*, suggested that school reformers at that time were interested
in advancing their own interests over those of others, particularly
immigrants and members of the working class.[4] It was this work that
helped to set the stage in many respects for the other "radical revisionists" who took up similar themes in the years that followed.

More recently, however, school reform has become a national
preoccupation, and historians of education have revisited the question of self-conscious educational change undertaken for a variety of
purposes. Adam Nelson discusses some of this literature in his chapter on the growing federal role in education during the postwar era.
But the most influential study of school reform in recent years has
been David Tyack and Larry Cuban's book, *Tinkering Toward
Utopia*, a sweeping examination of the reform process itself across
much of the twentieth century. Taking cues from earlier studies,
Tyack and Cuban are balanced but critical in their appraisal of
reform efforts in the past and present. In the end they suggest that
planning change in educational systems is a task almost always
doomed to some degree of failure, simply because schools and other
institutions are comprised of human relationships that rarely are
taken into account in the minds of reformers. Other historians who
have recently examined school reform in similar fashion include
James Carl, David Angus, and Jeffrey Mirel.[5] Both the promise and
limitations of reform are major points of emphasis, lessons especially
apt for an age when educational reform once again seems to have
become a national preoccupation.

The foregoing topics, of course, are only the most prominent of
new interests among historians of education. Like most other fields
of research and scholarship, the history of American education is
undergoing continual change and renewal. Even as these topics gain
greater attention and develop into full-blown subfields of inquiry
and debate, new areas of research are sure to emerge as well. This,
perhaps more than anything else, is one of the great lessons to be
derived from this collection of essays on a changing field of historical scholarship. The only safe prediction about the future is that
change is bound to occur. Looking back at the past several decades

of historical writing about education in the United States, we have
every reason to expect that the history of tomorrow will teach us
even more about the past, and about ourselves as well.

NOTES

1. Margaret Szasz, *Education and the American Indian: The Road to Self-Determination, 1928–1973* (Albuquerque: University of New Mexico Press, 1974); David Wallace Adams, *Education for Extinction: American Indians and the Boarding-School Experience, 1875–1928* (Lawrence: University Press of Kansas, 1995). On American Indians writing about these questions, see Devon A. Mihesuah, *Natives and Academics: Researching and Writing about American Indians* (Lincoln: University of Nebraska Press, 1998), and Norman T. Oppelt, *The Tribally Controlled Indian College: The Beginnings of Self-Determination in American Indian Education* (Tsaile, AR: Navajo Community College Press, 1990).

2. Guadalupe San Miguel, Jr., *Let All of Them Take Heed: Mexican Americans and the Campaign for Educational Equality in Texas, 1910–1981* (Austin: University of Texas Press, 1987); Guadalupe San Miguel, Jr., *Brown, Not White: School Integration and the Chicano Movement in Houston* (College Station: Texas A&M University Press, 2001); George J. Sánchez, *Becoming Mexican American: Ethnicity, Culture, and Identity in Chicano Los Angeles, 1900–1945* (New York: Oxford University Press, 1993); Eileen H. Tamura, *Americanization, Acculturation, and Ethnic Identity: The Nisei Generation in Hawaii* (Urbana: University of Illinois Press, 1994); also see Rubén Donato, *The Other Struggle for Equal Schools: Mexican Americans during the Civil Rights Era* (Albany, NY: State University of New York Press, 1997); and Gilbert G. Gonzalez, *Chicano Education in the Era of Segregation* (Philadelphia: Balch Institute Press, 1990), as well as Victoria-Maria MacDonald, *Latino Education in the United States: A Narrated History from 1513–2000* (New York: Palgrave Macmillan, 2004).

3. Donald Warren, ed. *American Teachers: Histories of A Profession at Work* (New York: Macmillan, 1989); Larry Cuban, *How Teachers Taught: Constancy and Change in American Classrooms, 1890–1980* (New York: Longman, 1984); Kate Rousmaniere, *City Teachers: Teaching and School Reform in Historical Perspective* (New York: Teachers College Press, 1997); Kathleen Weiler, *Country Schoolwomen: Teaching in Rural California, 1850–1950* (Stanford, CA: Stanford University Press, 1998); James W. Fraser, *Preparing America's Teachers: A History* (New York: Teachers College Press, 2006).

4. Lawrence Cremin, *The Transformation of the School; Progressivism in American Education, 1876–1957.* (New York: Knopf, 1961); Michael B. Katz, *The Irony of Early School Reform; Educational Innovation in Mid-Nineteenth Century Massachusetts* (Cambridge, MA: Harvard University Press, 1968)

5. David Tyack and Larry Cuban, *Tinkering Toward Utopia: A Century of Public School Reform* (Cambridge, MA: Harvard University Press, 1995); Jim Carl, "Harold Washington and Chicago's Schools Between Civil Rights and the Decline of the New Deal Consensus, 1955–1987," *History of Education Quarterly* 41 (Fall 2001): 311–43; David L. Angus and Jeffrey E. Mirel, *The Failed Promise of the American High School, 1890–1995* (New York: Teachers College Press, 1999).

INDEX